Advanced Techniques in Biological Electron Microscopy II

Specific Ultrastructural Probes

Edited by

J. K. Koehler

With Contributions by

S. S. Brown R. W. Davis P. Echlin J. Ferguson
S. W. Hui J. K. Koehler J. A. Lake
G. L. Nicolson D. F. Parsons W. D. Perkins
J.-P. Revel

With 105 Figures

Springer-Verlag Berlin Heidelberg New York 1978

James K. Koehler, Ph. D.
University of Washington
Department of Biological Structures
School of Medicine
Seattle, WA 98195/USA

The cover illustration shows a rat lymphocyte from bone marrow sequentially labeled with tritiated uridine (nuclear silver grains) and hemocyanin conjugated to anti Ig (hemocyanin on cell surface, *arrows*).

ISBN-13: 978-3-642-66811-1 e-ISBN-13: 978-3-642-66809-8
DOI: 10.1007/978-3-642-66809-8

Softcover reprint of the hardcover 1st edition 1978

Typesetting, offsetprinting and bookbinding: Konrad Triltsch, Graphischer Betrieb, Würzburg
2131/3130-543210

Preface

The use of the term "advanced" in the title of this book is somewhat arbitrary and very much relative with respect to time. Many techniques which were considered at the "cutting edge" of ultrastructural methodology just a few years ago are now routinely used in numerous laboratories. One could cite freeze-fracture, cryothin sectioning, or indeed most of the field of scanning electron microscopy as concrete examples. Thus the use of the term "advanced techniques" must be interpreted with regard to the present state of the art, and is useful only in informing the potential reader that this volume is not a primer to be used as an initial introduction into basic biological electron microscopy. Many excellent volumes have filled that niche in the past few years, and it is not intended that this modest book be a complete compendium of the field. Furthermore, any limited selection of papers on advanced techniques necessarily reflects the preferences and arbitrary whims of the editor, thereby excluding many equally important procedures which the knowledgeable reader will readily identify.

The first volume of this series appeared approximately five years ago and illustrated techniques which were thought to represent advanced and yet basically morphological methods for gaining increased ultrastructural information from biological specimens. The present volume, on the other hand, stresses techniques which provide specific physicochemical data on the specimens in addition to the structural information. The future importance of fine structural investigations would seem to be strongly dependent on our ability to adopt such methods to help answer some of the outstanding questions in cell biology.

Three of the contributions of this volume deal with the use of various surface probes having specific affinities for cell surface molecules. The utilization of labeled lectins to explore cell surfaces has grown explosively during the past few years, and is discussed in the chapter by G. L. NICOLSON. Antibody labels have also become a very powerful specific probe of surface activities and are treated in the chapter by W. D. PERKINS and J. K. KOEHLER. S. S. BROWN and J. P. REVEL deal with the use of these and other types of labeled probes in the scanning electron microscope. The use of these methods has already contributed some very important new information to improve our concepts of antigen processing, cell fusion, and exocytosis, to mention just a few examples. The area of scanning electron microscopy is further represented by the chapter on low temperature preparations contributed by P. ECH-

LIN. The localization of highly labile substances in biological materials continues to be a nagging problem and considerable promise for a solution seems to lie in such cryo techniques. Another procedure that will be useful for such investigations involves the use of hydrated specimens which is discussed from the point of view of electron diffraction as well as electron microscopic studies by S. W. HUI and D. PARSONS. Finally, in the area of nucleoprotein fine structure research, two of the most elegant procedures are presented in chapters by J. LAKE, J. FERGUSON and R. W. DAVIS. The first of these is an account of the detailed structure of the ribosome as determined from nascent antibody labeling studies, and the second describes the use of heteroduplex analysis in genetic mapping research.

My sincere appreciation is extended to the authors of these chapters for their time-consuming efforts and patience, to the staff of the Springer-Verlag for their dedication to the production of the highest quality scientific publications, and to Ms. DORIS RINGER for help in the editorial processing of the manuscripts.

Seattle, February 1978 JAMES K. KOEHLER

Contents

Cell Surface Labeling for the Scanning Electron Microscope

S. S. BROWN and J.-P. REVEL (With 3 Figures)

Low-Temperature Biological Scanning Electron Microscopy
P. ECHLIN (With 18 Figures)

Quantitative Electron Microscopy of Nucleic Acids
J. FERGUSON and R. W. DAVIS (With 26 Figures)

Electron Microscopy of Specific Proteins: Three-Dimensional Mapping of Ribosomal Proteins Using Antibody Labels

J. A. LAKE (With 32 Figures)

Electron Microscopy and Electron Diffraction Studies on Hydrated Membranes

S. W. HUI and D. F. PARSONS (With 11 Figures)

Contributors

BROWN, SUSAN S., Department of Structural Biology, Stanford University School of Medicine, Stanford CA 94305, USA

DAVIS, RONALD W., Department of Biochemistry, Stanford University School of Medicine, Stanford CA 94305, USA

ECHLIN, PATRICK, The Botany School, Downing Street, Cambridge, CB2 3EA, Great Britain

FERGUSON, JILL, Department of Biochemistry, University of Washington, Seattle, WA 98195, USA

HUI, SEK-WEN, Electron Optics Laboratory, Biophysics Department, Roswell Park Memorial Institute, Elm Street, Buffalo, NY 14263, USA

KOEHLER, JAMES K., University of Washington, Department of Biological Structures, Seattle, WA 98195, USA

LAKE, JAMES A., Molecular Biology Institute and Department of Biology, University of California, Los Angeles, CA 90024, USA

NICOLSON, GARTH L., Department of Developmental and Cell Biology, University of California, Irvine, CA 92717, USA

PARSONS, DONALD F., Electron Optics Lab., Division of Laboratories and Research, New York State Department of Health, Empire State Plaza, Albany, NY 12201, USA

PERKINS, WILLIAM D., Department of Biological Structure, University of Washington, School of Medicine, Seattle, WA 98195, USA

REVEL, JEAN-PAUL, Division of Biology 156-29, California Institute of Technology, Pasadena, CA 91125, USA

Ultrastructural Localization of Lectin Receptors

G. L. NICOLSON

A. Introduction

Lectins, proteins, or glycoproteins that bind to specific sequence(s) in oligosaccharides, have been extensiveley used in cell biology to examine certain aspects of cell surface and membrane structure and function (reviews: LIS and SHARON, 1973; NICOLSON, 1974 c; 1975; SHARON and LIS, 1975). The classical use for lectins is cell agglutination, and agglutinability has been found to vary with antigenic type, state of infection, growth (cell density and contact), stage of the cell cycle, during embryonic development, after oncogenic transformation, etc. (MÄKELÄ, 1957; LIS and SHARON, 1973; NICOLSON, 1974 c). In addition to their assets as cell oligosaccharide structural probes, lectins have found extensive use in membrane biology in studying the number, distribution, turnover, display and mobility, and biogenesis of membrane glyco-components. Quantitative lectin binding (using ^{125}I-, ^{3}H-, or ^{63}Ni-labels) has been successfully used to measure the number of lectin receptors, their affinities, and kinetics of lectin association (and dissociation), and fluorescent lectins have been utilized for low-resolution lectin receptor localization at the level of light microscopy. The variety of ultrastructural techniques for lectin receptor localization utilizing electron microscopy will be discussed further.

Many lectins possess important biological properties. Lectins can affect cell transport systems (ISSELBACHER, 1972; CZECH and LYNN, 1973), mimic hormone action (CUATRECASAS and TELL, 1973; CZECH and LYNN, 1973), stimulate mitosis (MÖLLER, 1970; POWELL and LEON, 1970; JANOSSY and GREAVES, 1971; ANDERSSON et al., 1972) and platelet release (MAJERUS and BRODIE, 1972), secretion of migration factors (SCHWARTZ et al., 1970), and histamine (SIRAGANIAN and SIRAGANIAN, 1974) as well as other processes (NICOLSON, 1974 c). They have been found to inhibit cell growth (DENT, 1971; INBAR et al., 1972; RALPH and NAKOINZ, 1973), movement (FRIBERG et al., 1972), phagocytosis (BERLIN, 1972), delayed hypersensitivity (LEON and SCHWARTZ, 1969), allograph rejection (MARKOWITZ et al., 1969) and fertilization (OIKAWA et al., 1973; 1974).

The number of different kinds of lectins available to the investigator is tremendous. For example, MÄKELÄ (1957) listed over 260 in his review and

TOMS and WESTERN (1971) compiled data for over 500 different lectins. However, most of these lectins have not been purified and characterized, so their use in electron-microscopic localization experiments is minimal. The most suitable lectins for electron microscopy and some of their properties are listed in Table 1. It is apparent from the diversity of different lectin-binding specificities that a variety of highly specific, but different cell surface sites can be probed by the use of a few lectins.

Several techniques for lectin-binding site localization are presented here along with their respective advantages and disadvantages, and simplified methods of synthesis of the ultrastructural probes will be detailed. The advice and help of many experts has been used to prepare this chapter, and their assistance is gratefully acknowledged (see Acknowledgements).

B. Purification of Lectins

Most lectins are obtained from plant or other sources as complex mixtures of proteins, lipids, and other contaminants. Before they can be used for electron-microscopic localization studies, it is advisable (or absolutely necessary in some cases) to purify them further by conventional or affinity techniques. Although the literature is rife with conventional lectin purification schemes using salt precipitation, gel chromatography, ion exchange chromatography, preparative electrophoresis, and others, the emphasis here is on affinity purification of lectins. Affinity purification procedures have been developed for most of the lectins in Table 1, and some of the more widely used affinity chromatography techniques are listed in Table 2. We have found for our own studies that affinity purified lectins yield lower backgrounds and greater binding specificity than all but those most highly purified by conventional procedures.

Affinity purified lectins can be radiolabeled before the ultrastructural probes are constructed. Although not done routinely, this serves three purposes: (1) the kinetics and extent of labeling can be determined independently on a large number of cells; (2) any loss of lectin label during the electron-microscopic preparation steps can be monitored; and (3) the ratio of lectin to ultrastructural marker (ferritin, peroxidase, hemocyanin, etc.) can be determined. A list of suitable isotope-labeling techniques for lectins has been presented in Table 2 of SHARON and LIS (1975), the most popular radiolabels being ^{125}I and ^{3}H.

C. Purification of Markers

Various ultrastructural markers have been used to identify lectins bound to their target sites. Some of these markers are covalently linked to lectins

Table 1. Some lectins suitable for electron microscopy

Origin	Common name or abbreviation	Mitogenic activity	Approximate molecular weight	Subunits	Blood group specific	Inhibitory saccharides	Key references
Canavalia einsformis (jack bean)	Con A (Concanavalin A)	+	55,000	2	—	α-*D*-Man, α-*D*-Glc	SUMNER and HOWELL (1936); SO and GOLDSTEIN (1967 a, b); AGRAWAL and GOLDSTEIN (1967, 1968); PORETZ and GOLDSTEIN (1968, 1970)
Dolichos biflorus (horse gram)	DBA (Dolichos biflorus agglutinin)	–	135,000	4	A	α-*D*-GalNAc	BIRD (1951); ETZLER and KABAT (1970); FONT et al. (1971)
Glycine max. (soybean)	SBA (Soybean agglutinin)	–[a]	110,000	2	—	*D*-GalNAc, *D*-Gal	WADA et al. (1958); LIS et al. (1966, 1969, 1970); GORDON et al. (1972 a, b); LIS and SHARON (1972); CATSIMPOOLAS and MEYER (1969)
Helix pomatia (snail)	—	?	100,000	~6	A	α-*D*-GalNAc	HAMMARSTROM and KABAT (1969, 1971); UHLENBRUCK et al. (1970 b); ISHIYAMA and UHLENBRUCK (1972)

[a] SBA is reported to be mitogenic to neuraminidase-treated lymphocytes (NOVOGRODSKY and KATCHALSKI, 1971)

Table 1 (continued)

Origin	Common name or abbreviation	Mitogenic activity	Approximate molecular weight	Subunits	Blood group specific	Inhibitory saccharides	Key references
Lens culinaris (lentil)	LCA-A	+	I = 60,000	2	—	α-*D*-Man, α-*D*-Glc	HOWARD and SAGE (1969); TICHÁ et al. (1970); TOYOSHIMA et al. (1970); HOWARD et al. (1971); PAULOVÁ et al. (1971 b); YOUNG et al. (1971)
	LCA-B (Lens culinaris agglutinin)	?	II = 42,000	2	—	α-*D*-Man, α-*D*-Glc	
Limulus polyphemus (horseshoe crab)	—	?	400,000	18	—	Sialic acid	MARCHALONIS and EDELMAN (1968); NOWAK and BARONDES (1975)
Lotus tetragonolobus	Lotus A	+	I = 120,000	4	H(O)	α-*L*-Fuc	YARIV et al. (1967); KALB (1968); MARCHALONIS and EDELMAN (1968)
	Lotus B	?	II = 58,000	2	H(O)	α-*L*-Fuc	
	Lotus C	?	III = 120,000	4	H(O)	α-*L*-Fuc	
Phaseolus vulgaris (red kidney bean)	L-PHA	+	I = 140,000	4?	—	*D*-GalNAc,	RIGAS and OSGOOD (1955); COULET et al. (1956); ALLAN et al. (1969); RIGAS and HEAD (1969); DAHLGREN et al. (1970); ALLEN and CRUMPTON (1971, 1973); SCHUMACHER et al. (1971); MILLER et al. (1973)
	H-PHA (Phytohemagglutinin)	+	II = 140,000	4?	—	*D*-GalNAc	
Pisum sativum (pea)	Pea agglutinin	+	I = 54,000	2	—	*D*-Man, *D*-Glc	ENTLICHER et al. (1969, 1970); PAULOVÁ et al. (1971 a)
		+	II = 53,000	2	—	*D*-Man, *D*-Glc	

Table 1 (continued)

Origin	Common name or abbreviation	Mito-genic activity	Approximate molecular weight	Sub-units	Blood group specific	Inhibitory saccharides	Key references
Ricinus communis (castor bean)	RCA_I Ricin, RCA_{II} (Ricinus communis agglutinin)	– –	I=120,000 II= 60,000	4 2	— —	β-*D*-Gal, *D*-Gal, *D*-GalNAc	KABAT et al. (1947); DRYSDALE et al. (1968); WALDSCHMIDT-LEITZ and KELLER (1969, 1970); GÜRTLER and HORSTMANN (1973); NICOLSON and BLAUSTEIN (1972); NICOLSON et al. (1974); OLSNES et al. (1974)
Sophora japonica	—	?	133,000	—	A+B	β-*D*-GalNAc	MILLER and BOYD (1967); PORETZ and TIMBERLAKE (1972); PORETZ et al. (1974)
Triticum vulgaris (wheat germ)	WGA (Wheat germ agglutinin)	–	23,000	2		$(D\text{-GlcNAc})_2$, sialic acid	BURGER and GOLDBERG (1967); UHLENBRUCK et al. (1968, 1970 a); BURGER (1969); BIDDLE et al. (1970); LEVINE et al. (1972); ALLEN et al. (1973); GREENAWAY and LEVINE (1973)
Ulex europaeus	UEA_I UEA_{II} (Ulex europaeus agglutinin)	+ –	I – 170,000 II=170,000	— —	H(O) H(O)	*L*-Fuc, $(D\text{-GlcNAc})_2$	MATSUMOTO and OSAWA (1969, 1970)

Table 2. Affinity purification of some lectins

Lectin	Specificity	Affinity adsorbent	Elutant	References
Con A	α-*D*-glc α-*D*-Man	Sephadex G-50	*D*-glc or sucrose	Agrawal and Goldstein (1967)
			glycine-HCl, pH 2	Olson and Liener (1967)
DBA	*D*-galNAc	polyleucyl (A+H substance)	*D*-galNAc	Etzler and Kabat (1970)
		Con A-Sepharose	α-Me-*D*-Man	Goldstein et al. (1973)
LCA	α-*D*-glc α-*D*-man	Sephadex-G100	*D*-glc	Tichá et al. (1970); Howard et al. (1971); Young et al. (1971)
P. vulgaris PHA	*D*-galNAc	Thyroglobulin-Sepharose	Glycine-HCl, pH 3	Matsumoto and Osawa (1972)
		Con A-Sepharose	α-Me-*D*-Man	Goldstein et al. (1973)
PSA	α-*D*-glc α-*D*-man	Sephadex-G150	Glycine-HCl, pH 2	Entlicher et al. (1970); Trowbridge (1974)
RCA_I	β-*D*-gal	Sepharose 4B	*D*-gal	Tomita et al. (1972)
		Biogel A 1.5 m	*D*-gal or lactose	Nicolson and Blaustein (1972); Nicolson et al. (1974)
SBA	*D*-gal *D*-galNAc	N-ε-aminocaproyl-β-*D*-galactopyranodylamine-Sepharose	*D*-gal	Gordon et al. (1972 a+b)
		Con A-Sepharose	α-Me-*D*-man	Goldstein et al. (1973)
SJA	α-*D*-gal	Polyleucyl (A+H substance)	*D*-gal	Poretz et al. (1974)
UEA	*L*-fuc	*L*-fucose-starch	glycine-HCl, pH 4	Matsumoto and Osawa (1972)
		Tri-N-acetylchitotriose-starch	Glycine-HCl, pH 3	Matsumoto and Osawa (1972)
WGA	$(D\text{-glcNAc})_2$	Ovomucoid-Sepharose	Acetic acid	Marchesi (1972); Nicolson et al. (1975 a)
		N-ε-aminocaproyl-β-D-glucopyranosylamine-Sepharose		Lotan et al. (1973)
		Chitin-Sepharose		Bloch and Burger (1974)

while others are bound to target-attached lectins by virtue of the fact they possess the appropriate carbohydrate sequence(s) for lectin binding. Simplified schemes for some markers are given below.

I. Hemocyanin

Busycon canaliculata are usually obtained from the Marine Biological Laboratory (Woods Hole, Mass., U.S.A.), but we have used Pacific keyhole limpets as well. The shell is broken with a hammer in the region of the heart, or a sharp knife is thrust through the foot into the heart, and the animals are placed into large supported funnels. Hemolymph is allowed to drip into a beaker at room temperature, and the debris is removed by low-speed centrifugation (15,000 *g* for 10 min). The supernatant is centrifuged at 59,000 *g* (Beckman-Spinco rotor Model Ti50 at 30,000 rpm or equivalent) for 15 to 30 min at 4° C to concentrate the hemocyanin (to 50 – 70 μg/ml) without pelleting. The resulting hemocyanin solution is about 90% pure at this point, and it is further purified by gel chromatography on Sepharose-2B or Biogel-A 0.5 m. The material is usually stored sterile at 4° C. It should not be frozen.

II. Ferritin

Many commercial grades of ferritin are available. Only a few are ready for ultrastructural work such as electron-microscopic grade ferritin (Polysciences, Inc., Warrington, Penna.; Immuno-diagnostic Systems, Solana Beach, Calif.; Sigma Chemical, St. Louis, Mo.). All other grades must be further purified before they can be used as electron-microscopic markers (HSU, 1967). Be aware that several manufacturers sell lyophilized ferritin that is partially denatured. If ferritin is frozen or lyophilized, *it cannot be used for electron microscopic work.* Orders for ferritin should contain directions to ship air mail in temperature-controlled polyfoam containers to prevent accidental freezing during transit.

Ferritin (2X or 6X crystallied, but *not* EM grade) can be purified by the following scheme: crystallization in 5% cadmium sulfate, precipitation in 50% ammonium sulfate, and ultracentrifugation.

1. *Cadmium Sulfate Crystallization*

Ferritin is diluted to 10 mg/ml in 2% ammonium sulfate and 0.35 volumes of 20% $CdSO_4$ are slowly added with stirring. The solution is stored at 4° C until crystals form and settle (usually 0.5 – 2 days); longer storage results in loss by adherence to glass. The ferritin crystals are carefully pelleted by centrifugation at 5000 *g* for 40 min at 4° C and the supernatant is discarded. Crystalline ferritin is dissolved in a volume of 2% ammonium sulfate equivalent to the initial volume of diluted ferritin, and the solution is centrifuged at 10,000 *g* for 35 min to remove aggregated or denatured material. Lack of a pellet at this stage indicates good ferritin purity. 20% $CdSO_4$ (0.35 volumes) is slowly added with stirring to the ferritin solution, and the crystallization sequence is continued. Usually 2 to 4 additional crystallizations are required for commercial ferritin.

2. *Ammonium Sulfate Precipitation*

Crystalline ferritin from the final $CdSO_4$ crystallization is dissolved in 0.75 volumes (based on the initial volume of diluted ferritin) of 2% ammonium sulfate and an equal volume of saturated ammonium sulfate solution is added. Allow the half-saturated ammonium sulfate solution to stand for 1 h at 4° C. The suspension is centrifuged at 10,000 *g* for 5 to 10 min to pellet the precipitate. Decant and discard the supernatant and carefully dissolve the pellet in a volume of distilled water equal to the initial volume of diluted ferritin. Add an equal volume of saturated ammonium sulfate and repeat the precipitation step. The final 50% saturated ammonium sulfate precipitate is centrifuged as above, dissolved in a small amount of distilled water and placed in a dialysis tube. Allow for a 2x increase in volume during dialysis and dialyze against *cold* (<16° C) running water overnight. Transfer to 0.05 M sodium phosphate buffer pH 7.5 and dialyze for several hours (overnight) at 4° C.

3. *Ultracentrifugation*

Centrifuge the purified ferritin from Section above at 80,000 *g* (35,000 rpm in a Beckman-Spinco Model 40 rotor) for 2 h at 4° C. Decant and discard the top 75 to 80% of the solution (containing mostly colorless apoferritin) and allow the ferritin pellet to dissolve slowly, overnight if necessary. The ferritin concentration can be determined by the method of LOWRY et al. (1951) or by absorbency at 440 nm by using the equation: ferritin concentration (mg/ml) = $A_{440nm}(0.65)$.

III. Peroxidase

Horseradish peroxidase (HRP) can be obtained commercially in fairly pure form (RZ : 3, Typ VI, Sigma Chemical, St. Louis, Mo., U.S.A.; Grade I, Boehringer, Mannheim, West Germany). There are many other types of peroxidases that can be substituted for HRP (such as the smaller microperoxidases), but most investigators routinely use HRP.

IV. Mannan-Iron Complex

The procedures for synthesis of an electron-dense mannan-iron complex follow those of FRANZ and ROTH (1975). Mannan is obtained from autoclaved yeast cells by extraction with citrate buffer and purified by the method of PEAT et al. (1961). The purified mannan has an average molecule weight of 30,000 to 70,000 and binds lectins recognizing *α-D*-Mann residues (Tables 1 and 2). To make the mannan-iron complex, mannan (2.5 mg) is dissolved in 5 ml water, and 5 ml of $FeCl_3$ solution (140 mg $FeCl_3$/ml) is added. This solution is slowly made alkaline by addition of 1*N* NaOH until pH 10.5 is obtained. At this point the solution is quickly heated to 80° C and maintained at this temperature for 30 min. It is next cooled, neutralized with HCl to pH 5.6 to 7.0, and the resulting precipitate removed by low-speed centrifugation. The mannan-iron complex in the supernatant is precipitated with 2 to 3 volumes of ethanol, pelleted by low-speed centrifugation, and dissolved in distilled water.

D. Synthesis of Probes and Labeling Techniques

In this section the synthesis of lectin-ultrastructural probes and procedures of labeling will be outlined. These instructions include descriptions from the literature and numerous details from investigators in the field.

I. Lectin-Hemocyanin

The distinctive size (mol wt $\sim 5-9 \times 10^6$) and cylindrical shape ($\sim 100 \times 350$ Å) of the hemocyanin molecule renders it identifiable in thin section (KARNOVSKY et al., 1972; UKENA et al., 1974) or in platinum-carbon replicas (SMITH and REVEL, 1972 and Table 3). It has also been used as

Table 3. Some uses of Con A-hemocyanin techniques

Con A receptors studies	Reference
Distribution on human and rat RBC, 3T3, SV3T3, and rabbit leukocytes	SMITH and REVEL (1972)
Dynamics on 3T3, SV3T3, trypsinized 3T3 cells	ROSENBLITH et al. (1973)
Effects of colchicine and cytochalasin B on Con A receptor dynamics on 3T3, SV3T3 cells	UKENA et al. (1974)
Capping of Con A receptors on polymorphonuclear leukocytes	RYAN et al. (1974)

a label for scanning electron microscopy (see chapter by BROWN and REVEL). The two-step labeling procedures of SMITH and REVEL (1972) have been used to label a variety of cells grown in situ or in suspension and the lectin-hemocyanin complexes observed by shadow cast techniques.

1. *Labeling Procedures*

Cells are grown on or attached to glass coverslips and are rinsed three times with phosphate-buffered saline, pH 7.4. The lectin solution (usually 0.5 ml of 100 – 200 μg/ml concanavalin A (Con A) or other suitable lectin from Table 1 in phosphate buffer) is applied to each coverslip. If the lectin solution is applied at 37° C, redistribution of lectin-binding sites occurs on most cells. Low-temperature (0° – 4° C) or prior aldehyde fixation (1% paraformaldehyde for 15 min at 22° C; 1% glutaraldehyde for 15 min at 4° C) dramatically reduces visible ligand-induced clustering, but higher concentrations of lectin ($\geqq$200 μg/ml) are necessary to obtain the same degree of labeling. After 10 min the coverslips are rinsed three times in buffer to remove excess lectin and a 0.5 ml solution of hemocyanin (0.5 – 1 mg/ml) in phosphate buffer is applied. The coverslips can be incubated at 0° to 4° C (using 1 mg/ml hemocyanin) or at 37° C (using 0.5 mg/ml hemocyanin) for 10 min. Next the coverslips are rinsed three times with phosphate-buffered saline, and the attached cells are fixed in phosphate-buffered 1% glutaraldehyde for 10 min at 22° C. Alternatively, the fixed, labeled cells can be incubated at various temperatures in buffer for additional periods of time before aldehyde fixation to demonstrate ligand-induced redistribution. The cells are finally fixed in buffered 2% osmium tetroxide for 10 min at 22° C and dehydrated in a graded series of ethanols. At this point the coverslips can be submerged into mixed isomers of amyl acetate for 1 min and air-dried according to SMITH and REVEL (1972) or critical point dried in a critical point drying apparatus.

2. *Platinum-Carbon Replicas*

By evaporating platinum at a 45° angle onto the coverslip, a shadow-cast replica can be made of the cell surface with three-dimensional qualities. Coverslips are placed in a vacuum evaporator, and the platinum electrode is set at an angle of 35 to 45° to the coverslip surface. When the vacuum reaches 1×10^{-5} Torr or better, a single burst of platinum evaporation (the amount of platinum evaporated must be determined by trial) results in a shadow cast replica that is strengthened by carbon evaporation using several 3 – 4 sec bursts with the electrode at a 90° angle to the specimen. Too light a carbon coating will not give sufficient strength to the replica, while an excess coating will

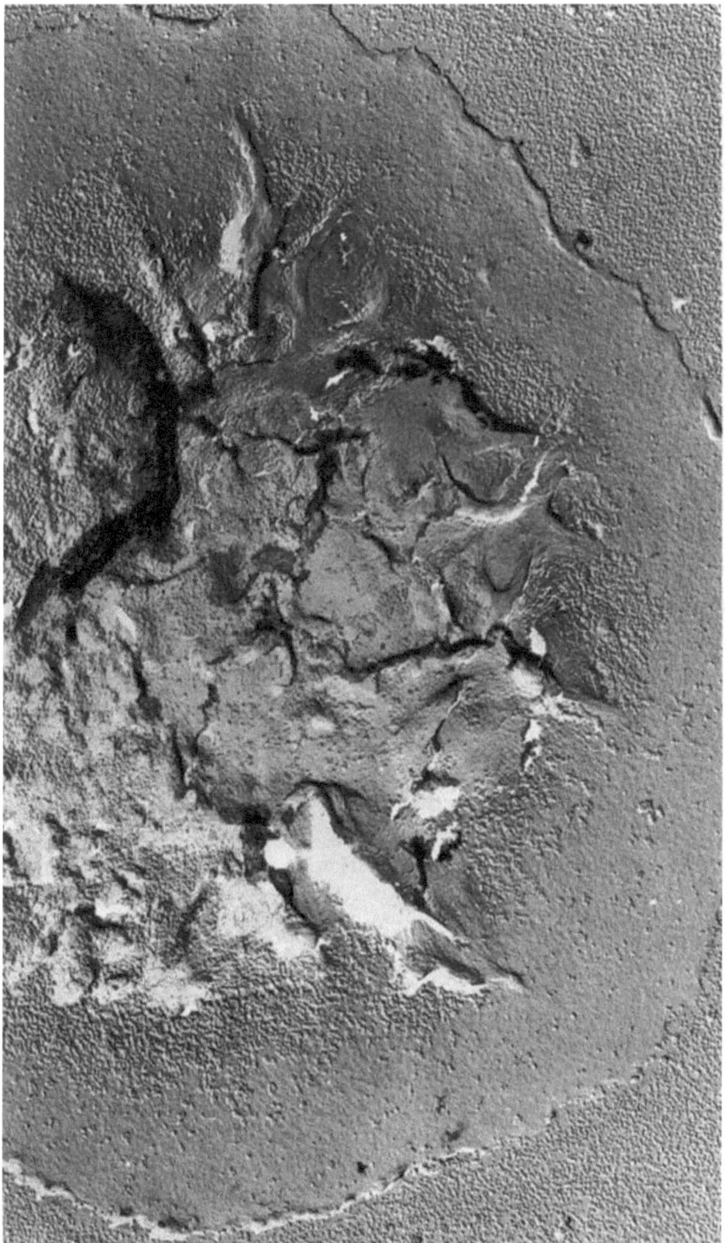

Fig. 1. Lectin-hemocyanin replica technique. A rabbit polymorphonuclear leukocyte was allowed to settle on a glass coverslip for 30 min and was labeled, unfixed, with Con A followed by hemocyanin, for 10 min each at 37° C. Lectin- and hemocyanin-induced redistribution has occurred. (From SMITH and REVEL, 1972; courtesy of DRS. S. SMITH-BROWN and J.-P. REVEL) × 19,200

give the replica a granular appearance. The replicas are cut into 2-mm squares using a diamond knife, and the coverslip is tilted (replica side up) into 48% hydrofluoric acid in a small Petri dish. Within seconds the replica squares loosen and are floated onto the acid surface where they can be picked up with wire loops after the surface tension has been reduced by dilution with distilled water. The replicas are transferred to a distilled water surface and then onto Clorox (sodium hypochlorite solution) for 1 h to remove the remaining biological material. After rinsing on distilled water, the replicas are finally picked up with 200-mesh uncoated grids.

SMITH and REVEL (1972) were able to study the topographic distribution of Con A-binding sites on a variety of cell types using the Con A-hemocyanin replica technique (Fig. 1). ROSENBLITH et al. (1973) extended the observations of SMITH and REVEL (1972) and studied the Con A receptor dynamics on 3T3 and SV3T3 cells. Prior formalin fixation or labeling at low temperature ($0° - 4°$ C) prevented ligand-induced redistribution of Con A-binding sites on SV3T3 cells. UKENA et al. (1974) used Con A-hemocyanin techniques in a comprehensive study on the effects of 2-deoxyglucose, colchicine, and cytochalasin B on surface Con A receptor distribution and mobility. These studies have the advantage of directly labeling cells on substrate to obtain a two-dimensional visualization of lectin-binding sites. Lectin receptors on microvilli and other surface structures can be easily identified. There are a few drawbacks to this technique. One appears to be the size of the hemocyanin marker ($\sim 100 \times 350$Å), which limits the resolution of localization. Another problem is quantitation of lectin-binding sites; several studies here indicated that the majority of the lectin-binding sites are *not* visualized (SMITH and REVEL, 1972). However, it is generally assumed that the lectin-binding sites that are visualized are representative. Finally, high backgrounds can be troublesome, and the conditions of labeling must be worked out for each system to minimize nonspecific adsorption of lectin and/or hemocyanin.

II. Lectin-Ferritin Conjugates

Various lectins habe been covalently coupled to ferritin for use as electron microscopic probes (Table 4). The most popular techniques use modification of the glutaraldehyde coupling procedures of AVRAMEAS (1969) by one- or two-step reactions.

1. *One-Step Glutaraldehyde Coupling*

The procedures generally follow those of NICOLSON and SINGER (1971; 1974) and NICOLSON (1971). Lectin and ferritin are dialyzed in 0.05 M sodi-

Table 4. Some lectin-ferritin conjugates and their uses

Lectin(s)	Biological material	Reference
Con A	3T3 and SV3T3 plasma membranes	NICOLSON (1971)
Con A	Human and rabbit RBC membranes	NICOLSON and SINGER (1971)
Con A	Normal lymphocyte	STOBO and ROSENTHAL (1972)
PHA	Human RBC ghosts	TILLACK et al. (1972)
Con A	3T3 and trypsinized-3T3	NICOLSON (1972 a)
Con A	Enzyme treated RBC	NICOLSON (1972 b)
Con A	Influenza virus and virus-infected cells	KLEIN and ADAMS (1972)
DBA	A-type human RBC	WILLIAMS and VOAK (1972)
RCA_I, Con A	Smooth and rough ER membranes form a myeloma	HIRANO et al. (1972)
Con A	3T3, SV3T3, trypsinized 3T3	DE PETRIS et al. (1973)
Helix pomatia	Ehrlich ascites	WAGNER and WAGNER (1973)
RCA_I	Influenza, VSV viruses	PENHOET et al. (1974)
Con A	Myelin and synaptosomes	MATUS et al. (1973)
RCA_I, Con A	Mouse lymphoma and fibroblast plasma membranes	NICOLSON and SINGER (1974)
Con A	3T3, SV3T3 in suspension	NICOLSON (1974 a)
Con A, RCA	Rat brain synaptosomes	BITTIGER and SCHNEBLI (1974); COTMAN and TAYLOR (1974)
RCA_I	Rabbit sperm	NICOLSON and YANAGIMACHI (1974)
Con A	3T3 cells in situ	NICOLSON (1975)
Con A	RNA-virus transformed cells	BARBARESE et al. (1973)
Con A	3T3, SV3T3 in suspension	RAFF et al. (1974)
RCA_I	Temperature-sensitive transformed hamster cells	NICOLSON et al. (1975 a)
Con A	Human RBC freeze-etch	PINTO DA SILVA and NICOLSON (1974)
RCA_{II}	3T3	NICOLSON (1974 b)
RCA_{II}	SV3T3, 3T3	NICOLSON et al. (1975 b)
Con A	Ehrlich ascites	ROTH et al. (1974)
RCA_{II}	Toxin sensitive and resistant lymphomas	NICOLSON et al. (1976)
Con A	Mouse lymphomas and fibroblasts	MARIKOVSKY et al. (1974)
RCA_I, WGA, Con A	Mouse, hamster, rat eggs	NICOLSON et al. (1975 c)
Con A	Human RBC — freeze-etch	BÄCHI and SCHNEBLI (1975)
Con A	Lymphocyte	YAHARA and EDELMAN (1975)

um phosphate buffer pH 6.8 – 7.2 [1] and mixed to form a solution containing 4 – 5% ferritin, 1.5 – 2% lectin, and 50 mM saccharide inhibitor in a small vial. The total volume is usually one ml, but smaller or larger preparations are possible. To this solution at room temperature (22° C) is slowly added 0.05 volumes of 0.6% purified [2] (neutral) glutaraldehyde, and the solution is mixed occasionally (or a small magnetic stirring bar can be added which rotates slowly at ~ 1 Hz) for 40 to 60 min or until a slight turbidity can be detected in strong light. One drop of 0.1 M ammonium chloride is added, and the reaction mixture is then centrifuged at 15,000 *g* for 10 min. The supernatant is carefully removed and applied to a 2.5 × 120 cm column of Sepharose 4B (Pharmacia) [3] or Biogel A-1.5 m^3 (Bio-Rad, Richmond, Calif.) equilibrated with the appropriate phosphate buffer. Fractions (1 ml) are collected when the visible ferritin band reach the bottom of the column. This procedure separates ferritin-oligomers, lectin-ferritin and ferritin from free lectin. The lectin-ferritin conjugate can be further purified by affinity chromatography using the appropriate affinity column (Table 2). Lectin-ferritin conjugates are usually stored at 4° C in a variety of buffers [1] and must be dialyzed back into 0.05 M Tris-HCl, pH 7.5, or loss of activity will result. Lectin-ferritin conjugates cannot be frozen and should be centrifuged occasionally to remove aggregates. Generally the conjugates are centrifuged immediately before use at 12,000 *g* for 10 min.

A more recent alternative procedure (NICOLSON et al., 1977) utilizes ^{125}I-labeled lectins. Affinity purified lectins are first radiolabeled with ^{125}I according to McFarlane (1958) by previously described procedures (NICOLSON, BLAUSTEIN and ETZLER, 1974). Ferritin-conjugates were synthesized and purified by a modification of the procedures of NICOLSON and SINGER (1974). ^{125}I-lectin was added to purified ferritin (Immuno-Diagnostics, Solana Beach, Calif.) in a ratio of 1 : 5 w/w in 0.2 M sodium chloride — 0.005 M sodium phosphate buffer, pH 7.2 (PBS) containing 0.2 M D-galactose. The reaction mixture was split into two equal portions and 20 μl of 1.3% EM grade glutaraldehyde in distilled water was added to one of the samples. Ten minutes later *and at each successive 10 min interval* 10 μl of 1.3% glutaraldehyde were added to both reaction mixtures. When the sample containing the additional 20 μl of glutaraldehyde displayed a slight turbidity under a strong beam of

---

[1] The pH depends on the lectin used; for example, 1 M sodium chloride and pH 6.8 are used for Con A to prevent aggregation, while 0.1 M sodium chloride and pH 7.2 – 7.4 are preferable for RCA_I, etc. The buffers used are generally those that have been used for lectin purification.

[2] Practical or commercial grade glutaraldehyde can be vacuum distilled using a small fractionating head or distilled without vacuum at 100° C. In both cases ample insulation is used to prevent excess temperatures that cause polymerization.

[3] Note: *β-D*-gal-specific lectins bind to agarose columns (see Tables 1 and 2).

visible light, the reactions in both samples were terminated by addition of 0.25 volumes of 1 M glycine in PBS. The samples were centrifuged at 27,000 xg for 15 min at 4° C and the small pellets discarded. The supernatants were pooled and centrifuged at 160,000 xg for 120 min onto a 0.5 ml cushion of Sepharose 4 B beads. After removal of the supernatants containing unconjugated ^{125}I-lectin, the pellet of ferritin and ferritin-^{125}I-lectin was gently resuspended in PBS and purified by affinity chromatography (see above).

2. *Two-Step Glutaraldehyde Method*

These procedures are an adaptation of the two-step coupling procedures of KISHIDA et al. (1975) and the procedures of OTTO et al. (1973) for antibody-ferritin conjugation.

To ferritin (10 mg/ml) in 0.05 M sodium phosphate buffer, pH 7.2, is added an excess amount of glutaraldehyde (one volume of 50% purified glutaraldehyde). The reaction is allowed to proceed with stirring (small magnetic bar at 1 Hz) for 20 to 30 min at 22° C or until a slight amount of turbidity is detected. The reaction mixture is quickly centrifuged at 15,000 *g* for 5 to 10 min at 0° C to remove aggregated ferritin. Unreacted glutaraldehyde is removed by rapid gel filtration at 4° C on a 2.5 × 120 cm column of Sephadex G-25 equilibrated with the phosphate buffer. Lectin solution containing 50 mM of the appropriate saccharide inhibitor is added, and the reaction mixture is concentrated by Amicon filtration (PM-30 membrane) or by Aquacide 11 (Calbiochem, San Diego, Calif.) at 4° C until the concentrations of ferritin and lectin approach 4 – 8 mg/ml and 2 – 4 mg/ml, respectively. Lectin-ferritin conjugates are then purified (see 1 above).

TAKAMIYA et al. (1974) compared the one-step and two-step glutaraldehyde conjugation procedures for antibodies and found the latter method to be slightly more effective with respect to resulting antibody activity of the ferritin conjugate. On the other hand, the one-step glutaraldehyde method for lectin-ferritin conjugation seems to be more reliable and offers a slightly better yield. One major advantage of the two-step method is the formation of fewer high molecular weight polymers of lectin and ferritin (author's unpublished observations).

3. *Labeling Procedures*

Since ferritin conjugates are covalently linked, a one-step labeling procedure is used. This can be made quantitative with radioisotope-lectin conjugates. For labeling intact cells the conjugate is added at various dilutions to several cell suspensions. Labeling temperature is important as temperatures above 5° C allow ligand-induced redistribution. The inherent distribution of lectin receptors can be obtained by labeling for short periods (1 – 10 min) at 0° C or by labeling formaldehyde-fixed or glutaraldehyde-fixed cells (as in INBAR et al., 1973; DE PETRIS et al., 1973). Mobility of lectin receptors and lectin-induced endocytosis has been investigated on certain cells by first labeling at low temperatures (0° – 4° C) and then elevating the temperature to 22° or 37° C (NICOLSON, 1972 a; 1974 b; NICOLSON and YANAGIMACHI, 1974; NICOLSON et al., 1975 a, b). For example, RCA_I-ferritin labeling at 0° C of a temperature-sensitive polyoma-transformed hamster BHK cell line (ts3-PyBHK) grown at the permissive (32° C) or nonpermissive (39° C) temperature for transformation results in dispersed distributions of RCA_I receptors (Fig. 2 a, b). Labeling at 0° C, washing, and then incubating at 22° C for 15 min results in relatively more lectin-induced clustering and endocytosis on ts3-Py-BHK (grown at 32° C) (Fig. 2 c) compared to ts3-PyBHK (grown at 39° C) cells (Fig. 2 d) (NICOLSON et al., 1975 a). After labeling, the cells are washed once or twice by low-speed centrifugation and fixed in 0.5 to 1.5% glutaraldehyde in phosphate-buffered saline, pH 7.4. Generally, cells are post-fixed in 1% osmium tetroxide, washed, and dehydrated for embedding. Controls usually contain 50 – 100 mM of the appropriate saccharide inhibitor in the labeling and wash solutions to specifically block lectin-ferritin binding.

Intact plasma membranes of certain cells can be directly mounted on thin films and observed after lectin-ferritin labeling (NICOLSON and SINGER, 1971). This procedure is straightforward for erythrocyte membranes (SINGER and NICOLSON, 1971; NICOLSON, 1972 a, b) (Fig. 3), viruses (PENHOET et al., 1974), etc., but it is more difficult to mount the plasma membranes of cells containing membrane-associated cytoskeletal elements (NICOLSON, 1971, 1972 a; BARBARESE et al., 1973; NICOLSON and SINGER, 1974; YAHARA and EDELMAN, 1975). A suspension of unfixed cells can be lysed on a hypotonic buffer surface by the following technique. A piece of parafilm or thin plastic containing several 5-mm diameter holes is placed over a 125-ml flask, and the flask is filled with buffer [4] until one additional drop in each hole results in a flat meniscus. One drop containing the cells (20 – 40% v/v) to be examined is added to each hole, and a few seconds later coated (formvar or collodion-carbon) 3-mm EM grids are used to pick up the cell membranes on the flat buffer meniscus. The grids are treated with 5% bovine serum albumin in buffer for 2 min at 22° C to prevent nonspecific lectin conjugate adsorption to the film, and the excess removed by a drop of buffer. One drop of the ferritin-conjugate is added (at 4°, 22°, or 37° C) and each grid is incubated for 2 to 5 min. At that time excess ferritin-lectin is removed by floating the grids serially for 10 to 20 sec each on 5 to 6 drops of buffer and finally on 2

[4] For unfixed cells use 10 MOs phosphate buffer, pH 7.4 (erythrocytes) and 250 M Os phosphate buffered saline, pH 7.4 (other cells). For fixed cells (fixed in 0.1 – 0.5% paraformaldehyde for 3 – 5 min at 22° C) use 10 MOs phosphate buffer, pH 7.2.

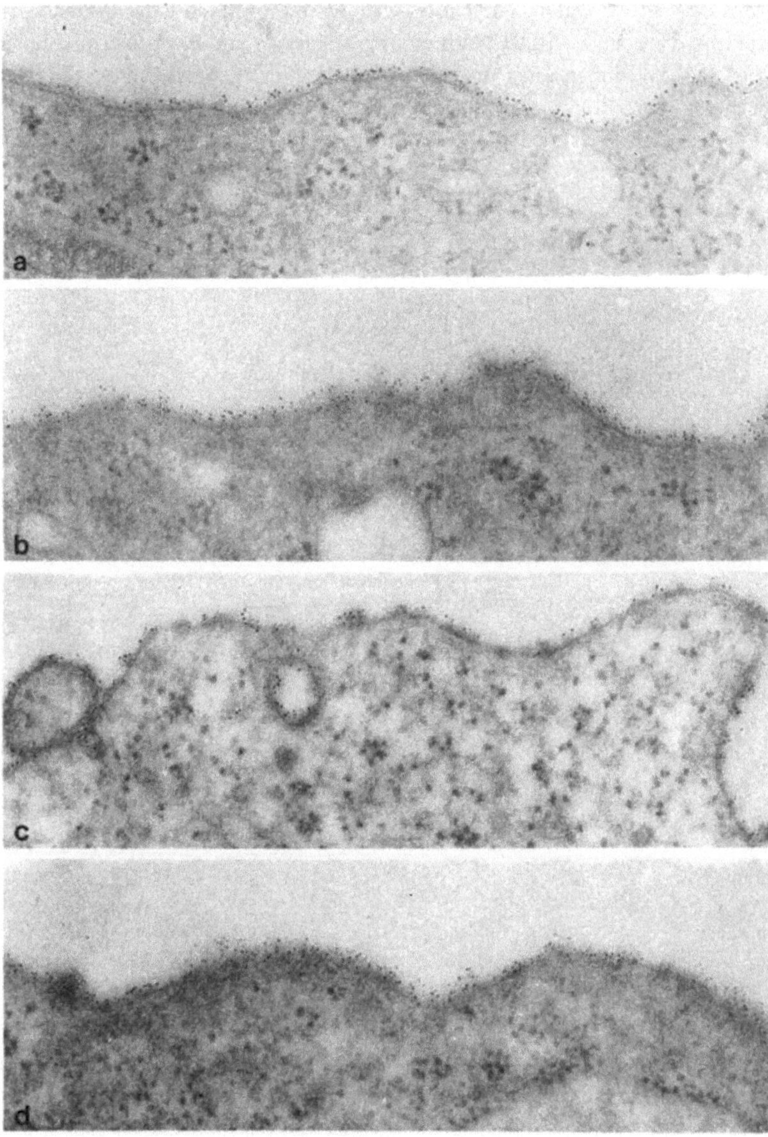

Fig. 2 a – d. Lectin-ferritin labeling to determine receptor dynamics. Temperature-sensitive polyoma-transformed hamster BHK cells were grown under permissive (32° C) and nonpermissive (39° C) conditions and labeled with ferritin-conjugated RCA_I. (a) ts3-Py-BHK cell grown at 32° C and labeled for 10 min at 0° C; (b) ts3-Py-BHK cell grown at 39° C and labeled for 10 min at 0° C; (c) same legend as in(a) except that the cells were washed after labeling at 0° C and incubated further at 22° C for 15 min; (d) same legend as in (b) except that cells were washed after labeling at 9° C and incubated at 22° C for 15 min. (From NICOLSON et al., 1975 a, by permission) × 45,000

drops of distilled water. The excess water is removed, and the grids are air dried or dried in a critical point drying apparatus and examined unstained (Fig. 3). Two-dimensional distributions of lectin receptors can be obtained by this method; however, the technique may be susceptible to potential artifacts, because cell lysis is required to obtain plasma membranes.

Fig. 3. Lectin-ferritin labeling of a mounted erythrocyte membrane. Human erythrocyte membranes were mounted on carbon-strengthened collodion-coated films and labeled 5 min with ferritin-conjugated LCA at 4° C. × 68,000

Recently lectin-ferritin labeling has been applied to freeze-etch electron microscopy of cells. In freeze-etch electron microscopy cells or their membranes are labeled with lectin-ferritin conjugates, washed, and frozen. The frozen samples are freeze-fractured at low temperature followed by controlled freeze-etching. The fracture planes of membranes are known to bisect their lipid bilayer structures (PINTO DA SILVA and BRANTON, 1970; TILLACK and MARCHESI, 1970) revealing particles embedded in the hydrophobic membrane regions. To reveal the outer surfaces of membranes where any lectins and their markers reside, etching or controlled sublimation is required (PINTO DA SILVA and BRANTON, 1970; TILLACK and MARCHESI, 1970). Intramembranous particles are known to be composed of glycoproteins and

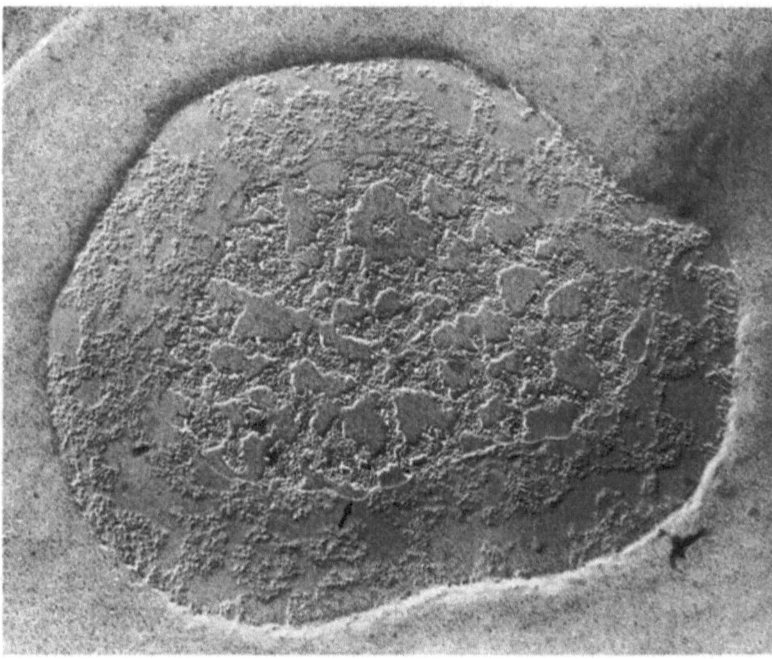

Fig. 4. Lectin-ferritin labeling followed by freeze-fracture and etching. Human erythrocyte membranes were incubated at pH 5.5 and labeled with ferritin-conjugated Con A and freeze-fractured followed by controlled freeze-etching. The border between the etched membrane surface and the internal fracture plane is indicated by an arrow. Con A binding sites are contiguous with the membrane-intercalated particles. (From PINTO DA SILVA and NICOLSON, 1974, by permission) × 34,000

proteins in at least erythrocyte membranes (PINTO DA SILVA et al., 1971; TILLACK et al., 1972; GRANT and McCONNELL, 1974; PINTO DA SILVA and NICOLSON, 1974). Using lectin-ferritin conjugates and freeze-fracture and etching techniques, the receptors for PHA (TILLACK et al., 1972), WGA (MARCHESI et al., 1972), and Con A (PINTO DA SILVA and NICOLSON, 1974) were found to be associated with the intramembranous particles in human erythrocytes (Fig. 4). In other plasma membranes this does not seem to be the case (KARNOVSKY et al., 1972; ROTH et al., 1974; PINTO DA

SILVA et al., 1975). Several other uses of lectin-ferritin conjugates are listed in Table 4.

Some investigators have had difficulty in obtaining good labeling intensities using ferritin conjugates compared to other labeling techniques (BRETTON et al., 1972 a), although in other studies quantitative data have been obtained with ferritin conjugates (NICOLSON et al., 1971). The size of the lectin-ferritin conjugate (~ 150 Å) may be important in determining the extent of labeling due to steric considerations; thus, certain types of receptors may not be accessible to labeling. The purity of ferritin is important in the construction of ferritin conjugates, and the conjugates, once synthesized, should be used within two weeks (author's unpublished observations) to alleviate possible aggregation, which has been reported to be a problem with some lectin-ferritin conjugates.

III. Lectin-Peroxidase Techniques

Peroxidase visualization utilizing diaminobenzidine (GRAHAM and KARNOVSKY, 1966) was first used by AVRAMEAS (1970) and BERNHARD and AVRAMEAS (1971) to make an electron-dense marker system for lectins. Since HRP is approximately 18% carbohydrate and contains lectin-accessible α-*D*-Mannoside residues, it can bind to the unoccupied valencies of cell surface-bound lectins such as Con A, LCA, PSA, and other lectins with similar binding specificities (Table 1). An alternative to this two-step labeling procedure has been developed by GONATUS and AVRAMEAS (1973) whereby peroxidase is covalently coupled to lectins such as Con A, RCA_I, WGA, etc. for use as one-step labels (Table 5).

1. *Two-Step Lectin-Peroxidase Labeling*

Cells (in Eagle's or Earle's media minus glucose or 25 mM Hepes without bicarbonate and glucose) are incubated with the appropriate lectin (usually 100 μg/ml) for 15 min at 4° or 22° C and washed three times with fresh media by slow-speed centrifugation at 4° C. Next the cells are incubated for 15 min at 4° or 22° C in excess media containing 50 μg/ml peroxidase and washed three times as before. Cells are fixed in 4% paraformaldehyde in 0.2 M sodium cacodylate phosphate buffer, pH 7.4 for 30 min at 22° C or in 1.25% glutaraldehyde in 0.1 M sodium phosphate buffer, pH 7.5 for 30 min at 4° C. Controls usually contain 0.1 M of the appropriate saccharide inhibitor in labeling and wash solutions. The fixed samples are treated for 15 to 30 min in the dark with a solution of 3,3′-diaminobenzidine hydrochloride (Sigma Chemical, St. Louis, Mo.; Fluka AG, Buch, Switzerland) (0.5 mg/ml) in 0.1 M Tris-HCl buffer, pH 7.4 containing 1 drop of 30% H_2O_2 per 5 ml of

Table 5. Some uses of lectin-peroxidase techniques

Lectin(s)	Biological material	Reference
Con A	Burkitt tumor and intestinal mucosa	Avrameas (1970); Bernhard and Avrameas (1971)
Con A	Fibroblast movement	Abercrombie et al. (1972)
Con A	Normal and polyoma-transformed cells	Martinez-Palomo et al. (1972)
Con A	Serum components adherent to substrate	Rowlatt and Wicker (1972)
Con A	Normal and SV40-transformed cells	Bretton et al. (1972 b)
WGA	Normal and transformed cells	Huet and Garrido (1972)
WGA, Con A	RNA virus-transformed cells	Francois et al. (1972)
Con A	Embryonic and adenovirus-transformed cells	Rowlatt et al. (1973)
Con A	EDTA and enzyme effects on hamster cells	Huet and Herzberg (1973)
Con A	Human lymphocytes	Barat and Avrameas (1973)
Con A	Normal and transformed cells	Torpier and Montagnier (1973)
WGA, Con A, RCA		Gonatus and Avrameas (1973)
LCA, Con A	Human and rabbit blood cells	Parmley et al. (1973)
Con A	Rat and human cells	Roth et al. (1973)
Con A	3T3 cells	Collard and Temmink (1974)
WGA, Con A	Normal and transformed hamster cells	Garrido et al. (1974)
Con A	Ehrlich ascites cells	Roth et al. (1974)
Con A	Efficiency of peroxidase labeling	Huet and Bernadac (1974)
Con A	Temperature effects on lectin binding and redistribution	Huet et al. (1974)
Con A	Normal liver and hepatoma cells	Roth (1974)
Con A	Human bone marrow cells	Ackerman and Waksal (1974)
LCA	Ehrlich ascites cells	Roth and Thoss (1974)
Con A	Rabbit sperm	Gordon et al. (1974)
LCA	Erythrocytes and tumor cells	Roth et al. (1975)
Con A	Endocytosis and dynamics on normal and transformed cells	Huet and Bernadac (1975)
Con A, RCA, LCA	Mouse lung cells	Roth (1975 a)
Con A	Normal and transformed cells	Collard et al. (1975)
LCA	Normal and transformed rat liver cells	Roth (1975 b)
Con A	Normal and transformed rat liver cells	Roth et al. (1975)
Con A, WGA	Hamster cells at various cell cycle stages	Garrido (1975)

incubation media (GRAHAM and KARNOVSKY, 1966). The cells are rinsed twice in media and postfixed in 2% osmium tetroxide in the phosphate buffer, dehydrated, and embedded in Araldite, "Epon" 812 or Spurr resin. Sections can be examined unstained or they can be counterstained briefly (2 – 5 min) with lead citrate.

The main advantages of the two-step lectin-peroxidase method are that it is uncomplicated, reliable, and labeling reagents do not have to be purified beyond that of commercial preparations. In addition, native unmodified lectin is used for the labeling experiments as opposed to lectin conjugates. However, there are some disadvantages. First, the reagents are used sequentially in two steps and are not covalently coupled. This means more washes and longer times between lectin, peroxidase, and fixation steps allowing some ligand induced redistribution to occur (ROWLATT et al., 1973; COLLARD and TEMMINK, 1974; GARRIDO et al., 1974; ROTH, 1974; HUET and BERNHARD, 1974). However, longer labeling times may be advantageous for revealing the dynamics of ligand-receptor complexes. Although labeling at low temperatures alleviates some redistribution, it can affect results significantly. Second, the use of noncovalently linked reagents in a two-step labeling sequence can result in some lectin and/or peroxidase loss by spontaneous release from cell surfaces (HUET and BERNADAC, 1975). Third, the lectin-peroxidase two-step technique is not quantititive, and specific lectin-receptor complexes may preferentially bind peroxidase, possibly in relation to the number of free lectin valencies at the cell surface (COLLARD and TEMMINK, 1974; HUET and BERNADAC, 1974). Fourth, proper controls without added lectin are necessary as endogenous cell peroxidases within certain types of cells will react in the diaminobenzidine reaction giving a false impression of lectin endocytosis, and finally, the resolution is limited by the size of the final electron-dense reaction product. On the other hand, the peroxidase product can be easily seen at low magnifications.

The dynamics of surface lectin receptors have been investigated using lectin-peroxidase two-step labeling procedures, including receptor mobility and redistribution (ROWLATT et al., 1973; COLLARD and TEMMINK, 1974; GARRIDO et al., 1974; HUET and BERNHARD, 1974; ROTH, 1974; ROTH et al., 1974), lectin-induced endocytosis (BERNHARD and AVRAMEAS, 1971; HUET and HERZBERG, 1973; HUET and BERNHARD, 1974; ROTH, 1974; HUET and BERNADAC, 1975; HUET and PEDROV, 1975), lectin and peroxidase release from cell surfaces (HUET and BERNADAC, 1975), and temperature effects on binding, redistribution, and endocytosis (HUET et al., 1974) (Table 5). An example of how lectin-peroxidase two-step labeling can be used to study Con A receptor dynamics is illustrated in Figure 5 (HUET and BERNHARD, 1974). Con A-peroxidase labeling of normal secondary hamster embryo cells at room temperature or after fixation yielded continous distributions of peroxidase product (c.f., Fig. 5 a). If the Con A-peroxidase-labeled unfixed cells were

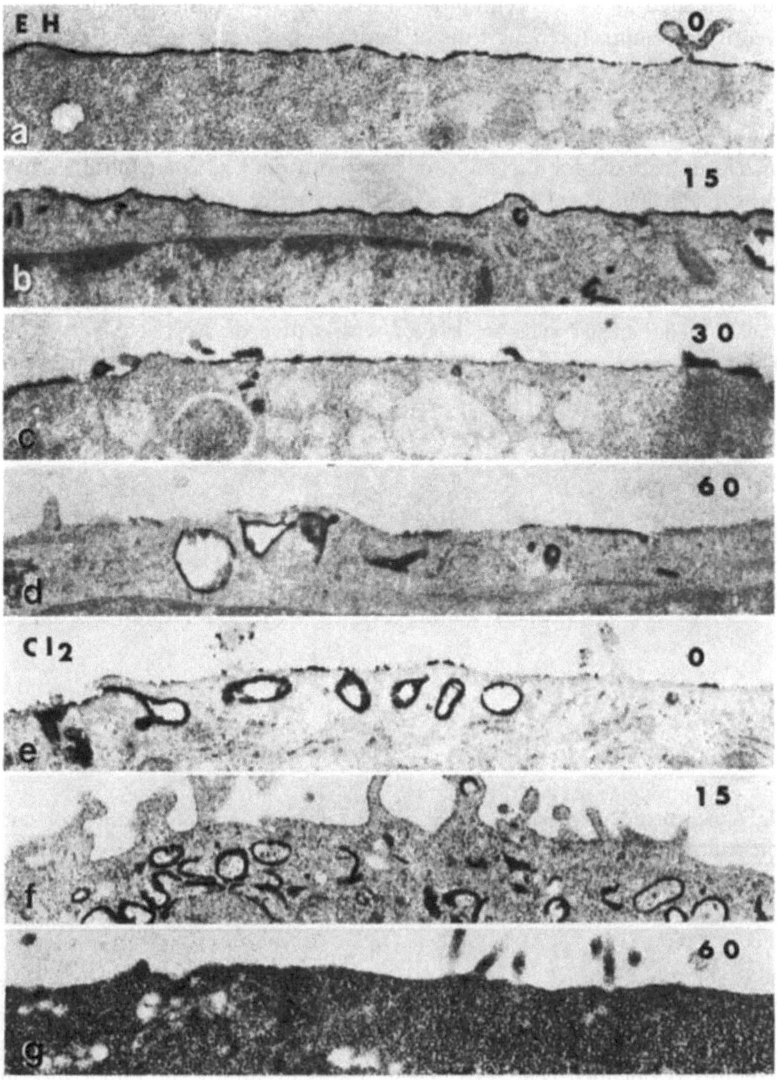

Fig. 5 a – g. Lectin-peroxidase techniques. Secondary hamster embryo and SV 40-transformed hamster cells labeled with Con A-peroxidase. (a) Control hamster embryo cells labeled with Con A, then peroxidase at 22° C, and immediately fixed. (b, c, d) Hamster embryo cells labeled at 22° C and postincubated at 37° C for 15, 30, and 60 min, respectively. (e) Control SV 40-transformed hamster cell labeled with Con A, then peroxidase at 22° C, and immediately fixed. (f, g) SV 40-transformed hamster cells labeled at 22° C and postincubated at 37° C for 15 and 60 min, respectively. (From HUET and BERNHARD, 1974, courtesy of Dr. C. HUET) × 14,000

postincubated at 37° C, lectin-induced redistribution occurred (Fig. 5 b, c), eventually leading to endocytosis of lectin-peroxidase complexes by 60 min (Fig. 5 d). When HUET and BERNHARD (1974) examined an SV 40-transformed hamster embryo cell line (Cl_2TSV_5), these events occurred even more rapidly than on the untransformed cell surface (Fig. 5 e – g), indicating a difference in the rate of mobility and endocytosis of Con A receptors on the transformed cell line. These results can be schematically shown in "glycograms" (Fig. 6) (HUET and BERNHARD, 1974). HUET et al. (1974) studied the temperature dependence of Con A labeling and ligand-induced redistribution and endocytosis on several untransformed and transformed cell lines. If SV 40-transformed hamster cells are labeled with native or acetylated Con A at 0° C, they show a continous labeling pattern (Fig. 7 a, b). Con A labeling at

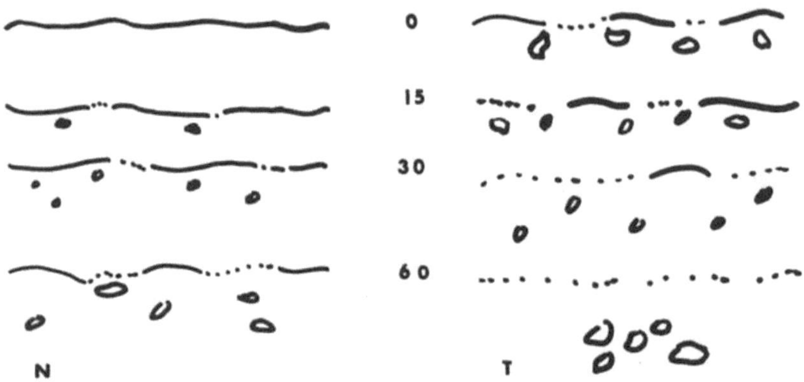

Fig. 6. Schematic 'glycograms' representing the different rates of redistribution and internalization of the cells labeled and shown in Figure 5. *N:* Normal secondary hamster embryo cells. *T:* SV-40 transformed hamster cells. (From HUET and BERNHARD, 1974, courtesy of Dr. C. HUET)

0° C for 2 h followed by peroxidase labeling at 37° C for 15 min results in dramatic rearrangement and endocytosis (Fig. 7 c), if the Con A-peroxidase labeling is performed as usual at 22° C, a patchy distribution of Con A receptors is obtained (Fig. 7 d). These results exemplify the importance in using the proper labeling and incubation conditions for determining the inherent distribution of cell surface lectin receptors, and they also indicate the utility of using a variety of labeling conditions to determine the mobility and modification of surface receptors.

2. *Single-Step Lectin-Peroxidase Labeling*

A one-step lectin-peroxidase labeling procedure has been developed by HUET and GARRIDO (1972) and GONATUS and AVRAMEAS (1973). Cova-

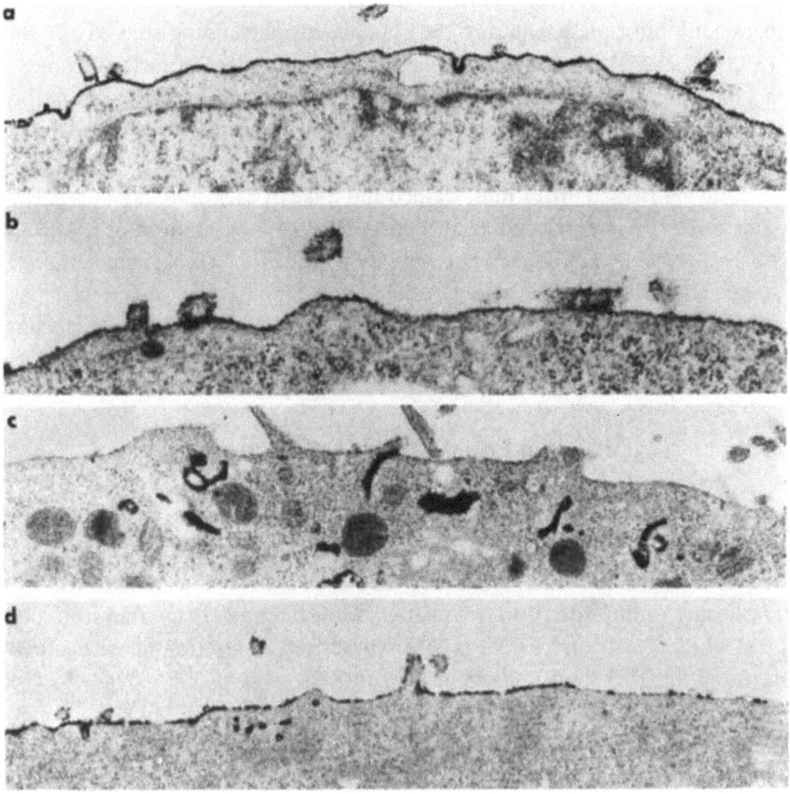

Fig. 7 a – d. Lectin-peroxidase techniques for determining receptor dynamics. (a) SV 40-transformed hamster cells labeled with Con A for 2 h then peroxidase at 0° C. (b) Same legend as in (a) except that acetylated (dimeric) Con A was used. (c) Cells labeled with Con A at 0° C for 2 h, then peroxidase for 15 min at 37° C. (d) Conventional Con A-peroxidase labeling at 25° C. (From HUET et al., 1974, courtesy of Dr. C. HUET) × 17,200

lently coupling the peroxidase to lectins eliminates the second peroxidase incubation. Procedures for coupling lectins and peroxidase are similar to those for lectin-ferritin coupling (Sect. C. II).

HUET and GARIDO's (1972) procedures are as follows: Lectin (1.5 mg/ml) and HRP (3 mg/ml) are mixed in 0.1 M sodium phosphate-buffered saline, pH 6.8. To this solution 0.03 ml of 1% purified glutaraldehyde is added, and the mixture is allowed to react for 1 to 3 h at 20° C. At that time the mixture is dialyzed against the same buffer containing additionally 2 mg/ml glycine. After two more dialysis steps at 4° C in the phosphate-buffered saline, the conjugate is centrifuged at 36,000 *g* for 20 min at 4° C to remove aggregated material. It is advisable to have 50 – 100 mM of the appropriate saccharide inhibitor present during the glutaraldehyde coupling step. HUET and GARRIDO (1972) recommend labeling at conjugate concentrations from 2.5 to 1200 μg/ml, and controls are run (1) in the presence of 0.5 M inhibitory saccharide and (2) with a peroxidase conjugate made without lectin being present during the coupling step.

Cells are labeled for 15 min, washed and fixed in 2.5% glutaraldehyde. Peroxidase activity is developed as in Section III.1., by the method of GRAHAM and KARNOVSKY (1966), and the cells are postfixed in 2% osmium tetroxide, washed and dehydrated as in Section III.1.

This method is advantageous to the two-step lectin-peroxidase labeling technique because the peroxidase incubation is eliminated.

IV. Lectin-Polysaccharide-Iron Complexes

A new two-step labeling procedure for the ultrastructural localization of Con A and lectins of similar saccharide specificity has been developed independently by MARTIN and SPICER (1974) and ROTH and FRANZ (1975). The former authors used Con A visualized by the binding of a dextran-iron complex, while FRANZ and ROTH (1975) synthesized an electron-dense mannan-iron complex that specifically binds to Con A.

1. Lectin-Dextran-Iron Complexes

Tissue or cells are fixed for 1 h at 22° C with phosphate-buffered 3% glutaraldehyde (pH 7.4) and washed three times (5 – 15 min each) in buffer. The fixed cells or tissue are reacted with Con A (1 mg/ml) at room temperature for 1 h and washed again three times for 5 to 15 min each 22° C. Dextran-iron complex (5 – 50 mg/ml; Imerferon, Lakeside Laboratories, Inc.) in phosphate-buffered saline is added and allowed to react with the Con A-labeled cells for 30 min at 22° C. Excess dextran-iron complex is removed by three washes as before, and the samples are postfixed in 1% osmium tetroxide in phosphate buffer for 1 h and dehydrated and embedded using standard procedures. The dextran-iron particles appear in a variety of sizes but are sufficiently electron dense to allow high resolution localization. The complex does not penetrate fixed cells, and controls without lectin are unlabeled, indicating absence of nonspecific binding to cell surfaces.

2. Lectin-Mannan-Iron Complexes

ROTH and FRANZ (1975) utilized a purified yeast mannan to make mannan-iron complexes (see Sect. C. IV) that bind to Con A, LCA, and other lectins of similar specificity. Their labeling procedures are as follows: Material to be labeled is fixed in phosphate-buffered 4% paraformaldehyde, pH 7.4 for 30 min at 22° C and washed three times in phosphate-buffered saline. A solution containing Con A (100 μg/ml) or LCA (1 mg/ml) is added and the incubation continued for 15 min at 22° C. After three washes in phosphate-buffered saline, a solution of yeast mannan-iron complex (mannan : iron = 85 : 15%) is added. After 15 min at 22° C the samples are washed three times and postfixed in 2% osmium tetroxide in 0.1 M sodium cacodylate buffer, pH 7.4 for 1 h at 4° C. Finally, the fixed samples are dehydrated, embedded, and examined unstained in thin section.

ROTH and FRANZ (1975) have used the lectin-mannan-iron technique to localize Con A and LCA receptors at the surfaces of mouse intestinal microvilli. The mannan-iron particles are quite visible in their unstained sections and occur in a variety of sizes. Controls containing α-Me-*D*-Man or mannan-iron complex alone were unlabeled indicating good specificity of labeling.

The advantage of this technique appears to be the ease in which the mannan-iron or dextran-iron label can be prepared. Disadvantages are those that plague all of the two-step labeling procedures involving polyvalent lectins and markers. Additionally, the variety of sizes and possible aggregation properties of the mannan-iron and dextran-iron complexes might be a problem where unambiguous localization of lectin receptors is necessary. Finally, these complexes react with lectins binding *D*-Man (Table 1), although other polysaccharide complexes should be available, depending on the sugar constituents of the polysaccharide complexes.

Acknowledgements. These studies were supported by a contract from the Tumor Immunology Program of the U.S. National Cancer Institute and grants from the U.S. Public Health Service and the Human Cell Biology Program of the U.S. National Science Foundation. The invaluable assistance and expert advice of Drs. S. AVRAMEAS, J. BORYSENKO, S. BROWN, C. HUET, M. KARNOVSKY, J.-P. REVEL, and J. ROTH are gratefully acknowledged.

References

Abercrombie, M., Heaysman, J. E. M., Pegrum, S. M.: Locomotion of fibroblasts in culture. V. Surface marking with concanavalin A. Exptl. Cell Res. 73, 536 – 539 (1972)

Ackerman, G. A., Waksal, S. D.: Ultrastructural localization of concanavalin A binding sites on the surface of differentiating hemopoietic cells. Cell Tiss. Res. **150**, 331 – 342 (1974)

Agrawal, B. B. L., Goldstein, I. J.: Protein-carbohydrate interaction. VI. Isolation of concanavalin A by specific adsorption on cross-linked dextran gels. Biochim. Biophys. Acta 147, 262 – 271 (1967)

Agrawal, B. B. L., Goldstein, I. J.: Protein-carbohydrate interaction. VII. Physical and chemical studies on concanavalin A, the hemagglutinin of the jack bean. Arch. Biochem. Biophys. **124**, 218 – 229 (1968)

Allan, D., Crumpton, M. J.: Fractionation of the phytohemagglutinin of *Phaseolus vulgaris* by polyacrylamide gel electrophoresis in sodium dodecyl sulphate. Biochem. Biophys. Res. Commun. **44**, 1143 – 1148 (1971)

Allan, D., Crumpton, M. J.: Phytohemagglutinin/lymphocyte interaction. Exptl. Cell Res. **78**, 271 – 278 (1973)

Allen, A. K., Neuberger, A., Sharon, N.: The purification, composition and specificity of wheat germ agglutinin. Biochem. J. **131**, 155 – 162 (1973)

Allen, L. W., Svenson, R. H., Yachnin, S.: Purification of mitogenic proteins derived from *Phaseolus vulgaris*. Isolation of potent and weak phytohemagglutinins possessing mitogenic activity. Proc. Natl. Acad. Sci. U.S. **63**, 334 – 341 (1969)

Andersson, J., Sjöberg, O., Möller, G.: Mitogens as probes for immunocyte activation and cellular cooperation. Transplant. Rev. **11**, 131 – 177 (1972)

Avrameas, S.: Coupling of enzymes to proteins with glutaraldehyde. Use of the conjugates for detection of antigens and antibodies. Immunochemistry **6**, 43 – 52 (1969)

Avrameas, S.: Emploi de la concanavaline-A pour l'isolement, la détection et la mesure des glycoprotéines et glucides extra- ou endo-cellulaires. C.R. Acad. Sci. (Paris) **270**, 2205 – 2208 (1970)

Bächi, T., Schnebli, H. P.: Reaction of lectins with human erythrocytes. II. Mapping of Con A receptors by freeze-etching electron microscopy. Exptl. Cell Res. **91**, 285 – 295 (1975)

Barat, N., Avrameas, S.: Surface and intracellular localization of concanavalin A in human lymphocytes. Exptl. Cell Res. **76**, 451 – 455 (1973)

Barbarese, E., Sauerwein, H., Simkins, H.: Alteration in the surface glycoproteins of chick erythrocytes following transformation with erythroblastosis strain R virus. J. Membrane Biol. **13**, 129 – 142 (1973)

Berlin, R. D.: Effect of concanavalin A on phagocytosis. Nature (New Biol) **235**, 44 – 45 (1972)

Bernhard, W., Avrameas, S.: Ultrastructural visualization of cellular carbohydrate components by means of concanavalin A. Exptl. Cell Res. **64**, 232 – 236 (1971)

Biddle, F., Cronin, A. P., Sanders, F. K.: The interaction between wheat germ agglutinin and receptors on normal and transformed cells and on erythrocytes. Cytobios **5**, 9 – 17 (1970)

Bird, G. W. G.: Specific agglutinating activity for human red blood corpuscles in extracts of *Dolichos biflorus*. Current Sci. (India) **20**, 298 – 299 (1951)

Bittiger, H., Schnebli, H.-P.: Binding of concanavalin A and ricin to synaptic junctions of rat brain. Nature (London) **249**, 370 – 371 (1974)

Bloch, R., Burger, M. M.: Purification of wheat germ agglutinin using affinity chromatography in chitin. Biochem. Biophys. Res. Commun. **58**, 13 – 19 (1974)

Bretton, R., Ternyck, T., Avrameas, S.: Comparison of peroxidase and ferritin labelling of cell surface antigens. Exptl. Cell Res. **71**, 145 – 155 (1972 a)

Bretton, R., Wicker, R., Bernhard, W.: Ultrastructural localization of concanavalin A receptors in normal and SV40-transformed hamster and rat cells. Int. J. Cancer **10**, 397 – 409(1972 b)

Burger, M. M.: A difference in the architecture of the surface membrane of normal and virally transformed cells. Proc. Natl. Acad. Sci. U.S. **62**, 994 – 1001 (1969)
Burger, M. M., Goldberg, A. R.: Identification of a tumor specific determinant on neoplastic cell surfaces. Proc. Natl. Acad. Sci. U.S. **57**, 359 – 366 (1967)
Catsimpoolas, N., Meyer, E. W.: Isolation of soybean hemagglutinin and demonstration of multiple forms by isoelectric focusing. Arch. Biochem. Biophys. **132**, 279 – 285 (1969)
Collard, J. G., Temmink, J. H. M.: Binding and cytochemical detection of cell-bound concanavalin A. Exptl. Cell Res. **86**, 81 – 86 (1974)
Collard, J. G., Temmink, J. H. M., Smets, L. A.: Cell cycle dependent agglutinability, distribution of concanavalin A binding sites and surface morphology of normal and transformed fibroblasts. In: Concanavalin A. Chowdhury, T. K., Weiss, A. K. (eds.). New York: Plenum Publishing Corp, pp. 221 – 244 (1975)
Cotman, C. W., Taylor, D.: Localization and characterization of concanavalin A receptors in the synaptic cleft. J. Cell Biol. **62**, 236 – 242 (1974)
Coulet, M., Cluzel, R., Bezon, M. J.: Procédé de purification de l'hémagglutinine de *Phaseolus vulgaris*. Transfusion **4**, 3 – 37 (1956)
Cuatrecasas, P., Tell, G. P. E.: Insulin-like activity of concanavalin A and wheat germ agglutinin — direct interactions with insulin receptors. Proc. Natl. Acad. Sci. U.S. **70**, 485 – 489 (1973)
Czech, M. P., Lynn, W. S.: Stimulation of glucose metabolism by lectins in isolated white fat cells. Biochim. Biophys. Acta **297**, 368 – 377 (1973)
Dahlgren, K., Porath, J., Lindahl-Kiessling, K.: On the purification of phytohemagglutinins from *Phaseolus vulgaris* seeds. Arch. Biochem. Biophys. **137**, 306 – 314 (1970)
Dent, P. B.: Inhibition by phytohemagglutinin of DNA synthesis in cultured mouse lymphomas. J. Nat. Cancer Inst. **46**, 763 – 773 (1971)
de Petris, S., Raff, M. D., Mallucci, L.: Ligand-induced redistribution of concanavalin A receptors on normal, trypsinized and transformed fibroblasts. Nature (New Biol.) **244**, 275 – 278 (1973)
Drysdale, R. G., Herrick, P. R., Franks, D.: The specificity of the haemagglutinin of the castor bean, *Ricinus communis*. Vox Sang. **15**, 194 – 202 (1968)
Entlicher, G., Koštíř, J. V., Kocourek, J.: Studies on phytohemagglutinins. III. Isolation and characterization of hemagglutinins from the pea (*Pisum sativum* L.). Biochim. Biophys. Acta **221**, 272 – 281 (1970)
Entlicher, G., Tichá, M., Koštíř, J. V., Kocourek, J.: Studies on phytohemagglutinins. II. Phytohemagglutinins of *Pisum sativum* L. and *Lens esculenta* moench: Specific interactions with carbohydrates. Experentia **25**, 17 – 19 (1969)
Etzler, M. E., Kabat, E. A.: Purification and characterization of a lectin (plant hemagglutinin) with blood group A specificity from *Dolichos biflorus*. Biochemistry **9**, 869 – 877 (1970)
Font, J., Leseney, A. M., Bourrillon, R.: Isolation and properties of an anti-A_1 phytohaemagglutinin from *Dolichos biflorus* seeds. Biochim. Biophys. Acta **243**, 434 – 446 (1971)
Francois, D., Vu-Van, T., Febvre, H., Haguenau, F.: Electron microscope study of the fixation of lectins labelled with Raifort peroxidase on human embryonic cells

transformed *in vitro* by the Rous sarcoma virus (RSV), Bryan strain. C.R. Acad. Sci. (Paris) **274**, 1981 – 1984 (1972)

Franz, H., Roth, J.: Preparation of carbohydrate-iron complexes as electron dense receptors for membrane bound lectins. Histochemistry **41**, 361 – 363 (1975)

Friberg, S., Jr., Golub, S. H., Lilliehöök, B., Cochran, A. J.: Assessment of concanavalin A reactivity to murine ascites tumours by inhibition of tumour cell migration. Exptl. Cell Res. **73**, 100 – 106 (1972)

Garrido, J.: Ultrastructural labeling of cell surface lectin receptors during the cell cycle. Exptl. Cell Res. **94**, 159 – 175 (1975)

Garrido, J., Burglen, M.-J., Samolyk, D., Wicker, R., Bernhard, W.: Ultrastructural comparison between the distribution of concanavalin A and wheat germ agglùtinin cell surface receptors of normal and transformed hamster and rat cell lines. Cancer Res. **34**, 230 – 243 (1974)

Goldstein, I. J., Reichert, C. M., Misaki, A., Gorin, P. A. J.: An "extension" of the carbohydrate binding specificity of concanavalin A. Biochim. Biophys. Acta **317**, 500 – 504 (1973)

Gonatus, N. K., Avrameas, S.: Detection of plasma membrane carbohydrates with lectin peroxidase conjugates. J. Cell Biol. **59**, 436 – 443 (1973)

Gordon, J. A., Blumberg, S., Lis, H., Sharon, N.: Purification of soy bean agglutinin by affinity chromatography. Meth. Enzymol. **28**, 365 – 368 (1972 a)

Gordon, J. A., Sharon, N., Lis, H.: Binding of soybean agglutinin by normal and trypsin-treated red blood cells. Biochim. Biophys. Acta **264**, 387 – 391 (1972 b)

Gordon, M., Dandekar, P. V., Bartoszewicz, W.: Ultrastructural localization of surface receptors for concanavalin A on rabbit spermatozoa. J. Reprod. Fert. **36**, 211 – 214 (1974)

Graham, R. C., Jr., Karnovsky, M. J.: The early stages of absorption of injected horseradish peroxidase in the proximal tubules of mouse kidney: Ultrastructural cytochemistry of a new technique. J. Histochem. Cytochem. **14**, 291 – 302 (1966)

Grant, C. W. M., McConnell, H. M.: Glycophorin in lipid bilayers. Proc. Natl. Acad. Sci. U.S. **71**, 4653 – 4657 (1974)

Greenaway, P. J., LeVine, D.: Binding of N-acetyl-neuraminic acid by wheat germ agglutinin. Nature (New Biol.) **241**, 191 – 192 (1973)

Gürtler, L. G., Horstmann, H. J.: Subunits of toxin and agglutinin of *Ricinus communis*. Biochim. Biophys. Acta **295**, 582 – 594 (1973)

Hammarström, S., Kabat, E. A.: Purification and characterization of a blood group A reactive hemagglutinin from the snail (*Helix pomatia*) and a study of its combining site. Biochemistry **8**, 2696 – 2705 (1969)

Hammarström, S., Kabat, E. A.: Studies on specificity and binding properties of the blood group A reactive haemagglutinin from *Helix pomatia*. Biochemistry **10**, 1684 – 1692 (1971)

Hirano, H., Parkhouse, B., Nicolson, G. L., Lennox, E. S., Singer, S. J.: Distribution of saccharide residues on membrane fragments from a myeloma-cell homogenate: Its implications for membrane biogenesis. Proc. Natl. Acad. Sci. U.S. **69**, 2945 – 2949 (1972)

Howard, I. K., Sage, H. J.: Isolation and characterization of a phytohemagglutinin from the lentil. Biochemistry **8**, 2436 – 2441 (1969)

Howard, I. K., Sage, H. J., Stein, M. D., Young, N. M., Leon, M. A., Dyckes, D. F.: Studies on a phytohemagglutinin from the lentil. II. Multiple forms of *Lens culinaris* hemagglutinin. J. Biol. Chem. **246**, 1590 – 1595 (1971)

Hsu, K. C.: Ferritin labeled antigens and antibodies. Meth. Immunol. Immunochem. **1**, 397 – 400 (1967)

Huet, C., Bernadac, A.: Peroxidase binding to cell-bound concanavalin A. Exptl. Cell Res. **89**, 429 – 431 (1974)

Huet, C., Bernadac, A.: Dynamic state of concanavalin A: Receptors interactions on fibroblasts surfaces. Biochim. Biophys. Acta **394**, 605 – 619 (1975)

Huet, C., Bernhard, W.: Differences in the surface mobility between normal and SV40-, polyoma- and adenovirus-transformed hamster cells. Int. J. Cancer **13**, 227 – 239 (1974)

Huet, C., Garrido, J.: Ultrastructural visualization of cell coat components by means of wheat germ agglutinin. Exptl. Cell Res. **75**, 523 – 527 (1972)

Huet, C., Herzberg, M.: Effects of enzymes and EDTA on ruthinium red and Con A labeling of the cell surface. J. Ultrastruct. Res. **42**, 186 – 199 (1973)

Huet, C., Lonchampt, M., Huet, M., Bernadac, A.: Temperature effects on concanavalin A molecule and on concanavalin A binding. Biochim. Biophys. Acta **365**, 28 – 39 (1974)

Huet, C., Petrou, J.: Interactions entre la concanavaline A et ses recepteurs a la surface cellulaire: Etude ultrastructurale. T. Micros. Biol. Cell. **23**, 213 – 228 (1975)

Inbar, M., Ben-Bassat, H., Sachs, L.: Inhibition of ascites tumor development by concanavalin A. Int. J. Cancer **9**, 143 – 149 (1972)

Inbar, M., Huet, C., Oseroff, A. R., Ben-Bassat, H., Sachs, L.: Inhibition of lectin agglutinability by fixation of the cell surface membrane. Biochim. Biophys. Acta **311**, 594 – 599 (1973)

Ishiyama, I., Uhlenbruck, G.: Further studies on the specificity of the anti-A agglutinin from *Helix pomatia*. Comp. Biochem. Physiol. **42 A**, 269 – 276 (1972)

Isselbacher, K. J.: Increased uptake of amino acids and 2-deoxy-D-glucose by virus-transformed cells in culture. Proc. Natl. Acad. Sci. U.S. **69**, 585 – 589 (1972)

Janossy, G., Greaves, M. F.: Lymphocyte activation. I. Response of T and B lymphocytes to phytomitogens. Clin. Exp. Immunol. **9**, 483 – 498 (1971)

Kabat, E. A., Heidelberger, M., Bezer, S.: A study of the purification and properties of ricin. J. Biol. Chem. **168**, 629 – 639 (1947)

Kalb, A. J.: The separation of three L-fucose binding proteins of *Lotus tetragonolobus*. Biochim. Biophys. Acta **168**, 532 – 536 (1968)

Karnovsky, M. J., Unanue, E. R., Leventhal, M.: Ligand-induced movement of lymphocyte membrane macromolecules. II. Mapping of surface moieties. J. Exp. Med. **136**, 907 – 930 (1972)

Kishida, Y., Olsen, B. R., Berg, R. A., Prockop, D. J.: Two improved methods for preparing ferritin-protein conjugates for electron microscopy. J. Cell Biol. **64**, 331 – 339 (1975)

Klein, P. A., Adams, W. R.: Location of ferritin-labeled concanavalin A binding to influenza virus and tumor cell surfaces. J. Virol. **10**, 844 – 854 (1972)

Leon, M. A., Schwartz, H. J.: Inhibition of delayed hypersensitivity to tuberculin by concanavalin A. Proc. Exp. Biol. Med. **131**, 735 – 736 (1969)

LeVine, D., Kaplan, M. J., Greenaway, P. J.: The purification and characterization of wheat-germ agglutinin. Biochem. J. **129**, 847 – 856 (1972)
Lis, H., Fridman, C., Sharon, N., Katchalski, E.: Multiple hemagglutinins in soybean. Arch. Biochem. Biophys. **117**, 301 – 309 (1966)
Lis, H., Sela, B., Sachs, L., Sharon, N.: Specific inhibition by *N*-acetyl-*D*-galactosamine of the interaction between soybean agglutinin and animal cell surface. Biochim. Biophys. Acta **211**, 582 – 585 (1970)
Lis, H., Sharon, N.: Soy bean (*Glycine max*) agglutinin. Meth. Enzymol. **28**, 360 – 368 (1972)
Lis, H., Sharon, N.: The biochemistry of plant lectins (phytohemagglutinins). Ann. Rev. Biochem. **43**, 541 – 574 (1973)
Lis, H., Sharon, N., Katchalski, E.: Identification of the carbohydrate-protein linking group in soybean agglutinin. Biochim. Biophys. Acta **192**, 364 – 366 (1969)
Lotan, R., Gussin, A. E. S., Lis, H., Sharon, N.: Purification of wheat germ agglutinin by affinity chromatography on a Sepharose-bound N-acetyl-glucosamine derivative. Biochem. Biophys. Res. Commun. **52**, 656 – 662 (1973)
Lowry, O. H., Rosebrough, N. J., Farr, S. L., Randall, R. J.: Protein measurement with folin phenol reagent. J. Biol. Chem. **193**, 265 – 275 (1951)
Majerus, P. W., Brodie, G. N.: The binding of phytohemagglutinins to human platelet plasma membranes. J. Biol. Chem. **247**, 4253 – 4273 (1972)
Mäkelä, O.: Studies in hemagglutinins of leguminosae seeds. Ann. Med. Exp. Fenn. **35** (suppl. 11), 1 – 133 (1957)
Marchalonis, J. J., Edelman, G. M.: Isolation and characterization of a hemagglutinin from *Limulus polyphemus*. J. Mol. Biol. **32**, 453 – 465 (1968)
Marchesi, V. T.: Purification of carbohydrate binding proteins: wheat germ (*Triticum vulgaris*) agglutinins. Meth. Enzymol. **28**, 354 – 356 (1972)
Marchesi, V. T., Tillack, T. W., Jackson, R. L., Segrest, J. P., Scott, R. E.: Chemical characterization and surface orientation of the major glycoprotein of the human erythrocyte membrane. Proc. Natl. Acad. Sci. U.S. **69**, 1445 – 1449 (1972)
Marikovsky, Y., Inbar, M., Danon, D., Sachs, L.: Distribution of surface charge and concanavalin A-binding sites on normal and malignant transformed cells Exptl. Cell Res. **89**, 359 – 367 (1974)
Markowitz, H., Person, D. A., Gitnick, G. L., Ritts, R. A.: Immunosuppressive activity of concanavalin A. Science **163**, 476 (1969)
Martin, B. J., Spicer, S. S.: Concanavalin A-iron dextran technique for staining cell surface mucosubstances. J. Histochem. Cytochem. **22**, 206 – 207 (1974)
Martinez-Palomo, A., Wicker, R., Bernhard, W.: Ultrastructural detection of concanavalin surface receptors in normal and in polyoma-transformed cells. Int. J. Cancer **9**, 676 – 683 (1972)
Matsumoto, I., Osawa, T.: Purification and characterization of an anti-H(O) phytohemagglutinin of *Ulex europaeus*. Biochim. Biophys. Acta **194**, 180 – 189 (1969)
Matsumoto, I., Osawa, T.: Purification and characterization of a *Cytisus*-type anti-H(O) phytohemagglutinin from *Ulex europaeus* seeds. Arch. Biochem. Biophys. **140**, 484 – 491 (1970)
Matsumoto, I., Osawa, T.: The specific purification of various carbohydrate-binding hemagglutinins. Biochem. Biophys. Res. Commun. **46**, 1810 – 1815 (1972)

Matus, A., de Petris, S., Raff, M. C.: Mobility of concanavalin A receptors in myelin and synaptic membranes. Nature (New Biol) **244**, 278 – 279 (1973)

McFarlane, A. S.: Efficient trace-labelling of proteins with iodine. Nature (London) **182**, 53 (1958)

Miller, J. B., Noyes, C., Heinrikson, R., Kingdon, H. S., Yachnin, S.: Phytohemagglutinin mitogenic proteins. Structural evidence for a family of isomitogenic proteins. J. Exp. Med. **138**, 939 – 951 (1973)

Miller, J. T., Boyd, W. C.: Immunochemical studies of lectins. I. Partial purification of the hemagglutinating protein of *Sophora japonica*. Vox Sang. **13**, 209 – 217 (1967)

Möller, G.: Immunocyte triggering. Cell. Immunol. **1**, 573 – 582 (1970)

Nicolson, G. L.: Difference in the topology of normal and tumor cell membranes as shown by different distributions of ferritin-conjugated concanavalin A on their surfaces. Nature (New Biol.) **233**, 244 – 246 (1971)

Nicolson, G. L.: Topography of cell membrane concanavalin A-sites modified by proteolysis. Nature (New Biol.) **239**, 193 – 197 (1972 a)

Nicolson, G. L.: Topological studies on the structure of cell membranes. In: Membrane Research. Fox, C. F. (ed.). New York: Academic Press, pp. 53 – 70 (1972 b)

Nicolson, G. L.: Factors influencing the dynamic display of lectin-binding sites on normal and transformed cell surfaces. In: Control of Proliferation in Animal Cell Surfaces. Clarkson, B., Baserga, B. (eds.). New York: Cold Spring Harbor Laboratory, pp. 251 – 270 (1974 a)

Nicolson, G. L.: Ultrastructural analysis of toxin binding and entry into mammalian cells. Nature (London) **251**, 628 – 630 (1974 b)

Nicolson, G. L.: The interactions of lectins with animal cell surfaces. Intern. Rev. Cytol. **39**, 89 – 190 (1974 c)

Nicolson, G. L.: Concanavalin A as a quantitative and ultrastructural probe for normal and neoplastic cell surfaces. In: Concanavalin A. Chowdhury, T. K., Weiss, A. K. (eds.). New York: Plenum Publishing Corp 1975, pp. 153 – 172

Nicolson, G. L., Blaustein, J.: The interaction of *Ricinus communis* agglutinin with normal and tumor cell surfaces. Biochim. Biophys. Acta **266**, 543 – 547 (1972)

Nicolson, G. L., Blaustein, J., Etzler, M. E.: Characterization of two plant lectins from *Ricinus communis* and their quantitative interaction with a murine lymphoma. Biochemistry **13**, 196 – 204 (1974)

Nicolson, G. L., Lacorbiere, M., Eckhart, W.: Qualitative and quantitative interactions of lectins with untreated and neuraminidase-treated normal, wild-type and temperature-sensitive polyoma-transformed fibroblasts. Biochemistry **14**, 172 – 179 (1975 a)

Nicolson, G. L., Lacorbiere, M., Hunter, T. R.: Mechanism of cell entry and toxicity of an affinity purified lectin from *Ricinus communis* and its differential effects on normal and virus-transformed fibroblasts. Cancer Res. **35**, 144 – 155 (1975 b)

Nicolson, G. L., Masouredis, S. P., Singer, S. J.: Quantitative two-dimensional ultrastructural distribution of $Rh_0(D)$ antigenic sites on human erythrocyte membranes. Proc. Natl. Acad. Sci. U.S. **68**, 1416 – 1420 (1971)

Nicolson, G. L., Robbins, J. C., Hyman, R. A.: Cell surface receptors and their dynamics on toxin-treated malignant cells. J. Supramol. Struc. 4, 15 – 26 (1976)

Nicolson, G. L., Singer, S. J.: Ferritin-conjugated plant agglutinins as specific saccharide stains for electron microscopy: Application to saccharides bound to cell membranes. Proc. Natl. Acad. Sci. U.S. **68**, 942 – 945 (1971)

Nicolson, G. L., Singer, S. J.: The distribution and asymmetry of saccharides on mammalian cell membrane surfaces utilizing ferritin-conjugated plant agglutinins as specific saccharide stains. J. Cell Biol. **60**, 236 – 248 (1974)

Nicolson, G. L., Smith, J. R., Hyman, R.: Dynamics of toxin and lectin receptors on a lymphoma cell line and its toxin-resistant variant using ferritin-conjugated, 125-iodine-labeled ligand. J. Cell Biol. (submitted, 1977)

Nicolson, G. L., Yanagimachi, R.: Mobility and the restriction of mobility of plasma membrane lectin-binding components. Science **184**, 1294 – 1296 (1974)

Nicolson, G. L., Yanagimachi, R., Yanagimachi, H.: Ultrastructural localization of lectin-binding sites on the zonae pellucidae and plasma membranes of mammalian eggs. J. Cell Biol. **66**, 263 – 274 (1975 c)

Novogrodsky, A., Katchalski, E.: Lymphocyte transformation induced by concanavalin A and its reversion by methyl-alpha-*D*-mannopyranoside. Biochim. Biophys. Acta **228**, 579 – 583 (1971)

Nowak, T. P., Barondes, S. H.: Agglutinin from *Limulus polyphemus:* purification with formalinized horse erythrocytes as the affinity adsorbent. Biochim. Biophys. Acta **393**, 115 – 123 (1975)

Oikawa, T., Nicolson, G. L., Yanagimachi, R.: Inhibition of hamster egg fertilization by phytoagglutinins. Exptl. Cell Res. **83**, 239 – 246 (1974)

Oikawa, T., Yanagimachi, R., Nicolson, G. L.: Wheat germ agglutinin blocks mammalian fertilization. Nature (London) **241**, 256 – 259 (1973)

Olsnes, S., Saltvedt, E., Pihl, A.: Isolation and comparison of galactose-binding lectins from *Abrus precatorius* and *Ricinus communis.* J. Biol. Chem. **249**, 803 – 810 (1974)

Olson, M. O. J., Liener, I. E.: The association and dissociation of concanavalin A, the phytohemagglutinin of the jack bean. Biochemistry **6**, 3801 – 3808 (1967)

Otto, H., Takamiya, H., Vogt, A.: A two-stage method for cross-linking antibody globulin to ferritin by glutaraldehyde. Antigen binding and agglutinating capacity. J. Immunol. Meth. **3**, 137 – 146 (1973)

Parmley, R. T., Martin, V. J., Spicer, S. S.: Staining of blood cell surfaces with a lectin-horseradish peroxidase method. J. Histochem. Cytochem. **21**, 912 – 922 (1973)

Paulová, M., Entlicher, G., Tichá, M., Koštíř, J. V., Kocourek, J.: Studies on phytohaemagglutinins. VII. Effect of Mn^{2+} and Ca^{2+} on haemagglutination and polysaccharide precipitation by phytohaemagglutination of *Pisum sativum.* Biochim. Biophys. Acta **237**, 513 – 518 (1971 a)

Paulová, M., Tichá, M., Entlicher, G., Koštíř, J. V., Kocourek, J.: Studies on phytohemagglutinins. IX. Metal content and activity of the hemagglutinin from the lentil (*Lens esculenta* moench). Biochim. Biophys. Acta **252**, 388 – 395 (1971 b)

Peat, S., Whelar, R., Edwards, T. E.: Polysaccharides of Baker's yeast. IV. Mannan. J. Chem. Soc. **29**, 29 – 34 (1961)

Penhoet, E., Olsen, C., Carlson, S., Lacorbiere, M., Nicolson, G. L.: Quantitative interaction of *Ricinus communis* agglutinin and concanavalin A with influenza and vesicular stomatitis viruses and virus-infected normal and polyoma-transformed cells. Biochemistry **13**, 3561 – 3566 (1974)

Pinto da Silva, P., Branton, D.: Membrane splitting in freeze-etching: Covalently bound ferritin as a membrane marker. J. Cell Biol. **45**, 598 – 605 (1970)

Pinto da Silva, P., Branton, D., Douglas, S. A.: Localization of A_1 antigen sites on human erythrocyte ghosts. Nature (London) **232**, 194 – 196 (1971)

Pinto da Silva, P., Martinez-Palomo, A., Gonzalez-Robles, A.: Membrane structure and surface coat of *Entamoeba histolytica*. J. Cell Biol. **64**, 538 – 550 (1975)

Pinto da Silva, P., Nicolson, G. L.: Freeze-etch localization of concanavalin A receptors to the membrane intercalated particles on human erythrocyte membranes. Biochim. Biophys. Acta **363**, 311 – 319 (1974)

Poretz, R. D., Goldstein, I. J.: The hydrophobic character of phenyl glycosidases and its relation to the binding of saccharides to concanavalin A. Arch. Biochem. Biophys. **125**, 1034 – 1036 (1968)

Poretz, R. D., Goldstein, I. J.: An examination of the topography of the saccharide binding sites of concanavalin A and of the forces involved in complexation. Biochemistry **9**, 2890 – 2896 (1970)

Poretz, R. D., Riss, H., Timberlake, J. W., Chien, S.: Purification and properties of the hemagglutinin from *Sophora japonica* seeds. Biochemistry **13**, 250 – 256 (1974)

Poretz, R. D., Timberlake, J. W.: Purification and properties of the *Sophora japonica* lectin. 164th Mtg. Am. Chem. Soc., New York, New York, September 1972. Abstract BIOL 163

Powell, A. E., Leon, M. A.: Reversible interaction of human lymphocytes with the mitogen concanavalin A. Exptl. Cell Res. **62**, 315 – 325 (1970)

Raff, M. C., de Petris, S., Mallucci, L.: Distribution and mobility of membrane macromolecules: Ligand-induced redistribution of concanavalin A receptors and its relationship to cell agglutination. In: Control of Proliferation in Animal Cell Surfaces. Clarkson, B., Baserga, R. (eds.). New York: Cold Spring Harbor Laboratory, pp. 271 – 282 (1974)

Ralph, P., Nakoinz, I.: Inhibitory effects of lectins and lymphocyte mitogens on murine lymphomas and myelomas. J. Nat. Cancer Inst. **51**, 883 – 890 (1973)

Rigas, D. A., Head, C.: The dissociation of phytohemagglutinin of *Phaseolus vulgaris* by 8.0 M urea and the separation of mitogenic from the erythroagglutinating activity. Biochem. Biophys. Res. Commun. **34**, 633 – 639 (1969)

Rigas, D. A., Osgood, E. E.: Purification and properties of the phytohemagglutinin of *Phaseolus vulgaris*. J. Biol. Chem. **212**, 607 – 615 (1955)

Rosenblith, J. Z., Ukena, T. E., Yin, H. H., Berlin, R. D., Karnovsky, M. J.: A comparative evaluation of the distribution of concanavalin A-binding sites on the surfaces of normal, virally-transformed, and protease-treated fibroblasts. Proc. Natl. Acad. Sci. U.S. **70**, 1625 – 1629 (1973)

Roth, J.: Distribution of concanavalin A receptors on normal rat liver cells and Zajdela ascites hepatoma cells. Int. J. Cancer **14**, 762 – 770 (1974)

Roth, J.: Electron microscopic demonstration of saccharide moieties in the hypophase of the alveolar surfactant system. Respir. Physiol. **23**, 325 – 335 (1975 a)

Roth, J.: *Lens culinaris* lectin receptors in the plasma membrane of rat liver cells: comparative electron microscopic studies on normal cells, on cells *in vivo* transformed by diethylnitrosamine and on Zajdela ascites hepatoma cells. Exptl. Path. **11**, 123 – 132 (1975 b)

Roth, J., Franz, H.: Ultrastructural detection of lectin receptors by cytochemical affinity reaction using mannan-iron complex. Histochemistry **41**, 365 – 368 (1975)

Roth, J., Meyer, H. W., Neupert, H., Bolck, F.: Concanavalin A binding sites in the plasma membrane of normal cells, spontaneously transformed cells and tumor cells as visualized by electron microscopy. Exp. Pathol. **8**, 19 – 30 (1973)

Roth, J., Neupert, G., Thoss, K.: Interaction of *Lens culinaris* lectin, concanavalin A, *Ricinus communis* agglutinin and wheat germ agglutinin with the cell surface of normal and transformed rat liver cells. Exp. Pathol. **10**, 309 – 317 (1975)

Roth, J., Thoss, K.: Light and electron microscopic demonstration of D-mannose and D-glucose like sites at the cell surface by means of the lectin from the *Lens culinaris*. Experientia **30**, 414 (1974)

Roth, J., Wagner, M., Meyer, H. W.: Topography of concanavalin A receptors on Ehrlich ascites tumour cells as demonstrated by thin sections and freeze etching. Pathol. Europ. **9**, 31 – 42 (1974)

Rowlatt, C., Wicker, R.: Concanavalin A binding material derived from a tissue culture medium. U. Ultrastruct. Res. **40**, 145 – 151 (1972)

Rowlatt, C., Wicker, R., Bernhard, W.: Ultrastructural distribution of concanavalin A receptors on hamster embryo and adenoviris tumour cell cultures. Int. J. Cancer **11**, 314 – 326 (1973)

Ryan, G. B., Borysenko, J. Z., Karnovsky, M. J.: Factors affecting the redistribution of surface-bound concanavalin A on human polymorphonuclear leukocytes. J. Cell Biol. **62**, 351 – 365 (1974)

Schumacher, K., Oerkermann, H., Uhlenbruck, G., Alzer, G., Hirschmann, W. D., Gross, R.: Isolation of the lymphoagglutinating fractions of phytohaemagglutinin (PHA) from *Phaseolus vulgaris* and characterization of their biological properties. Klin. Wschr. **49**, 286 – 288 (1971)

Schwartz, H. J., Pelley, R. P., Leon, M. A.: Release of a migration inhibitory factor from non-immune lymphoid cells by concanavalin A. Fed. Proc. **29**, 360 (1970)

Sharon, N., Lis, H.: Use of lectins for the study of membranes. Meth. Membrane Biol. **3**, 147 – 200 (1975)

Singer, S. J., Nicolson, G. L.: The structure and chemistry of mammalian cell membranes. Am. J. Pathol. **65**, 427 – 437 (1971)

Siraganian, P. A., Siraganian, R. P.: Basophil activation by concanavalin A: Characteristics of the reaction. J. Immunol. **112**, 2117 – 2125 (1974)

Smith, S. B., Revel, J.-P.: Mapping of concanavalin A binding sites on the surface of several cell types. Develop. Biol. **27**, 434 – 441 (1972)

So, L. L., Goldstein, I. J.: Protein-carbohydrate interaction. IX. Application of the quantitative hapten inhibition technique to polysaccharide-concanavalin A interaction. Some comments on the forces involved in concanavalin A-polysaccharide interaction. J. Immunol. **99**, 158 – 163 (1967 a)

So, L. L., Goldstein, I. J.: Protein-carbohydrate interaction. IV. Application of the quantitative precipitin method to polysaccharide-concanavalin A interaction. J. Biol. Chem. **242**, 1617 – 1622 (1967 b)

Stobo, J. D., Rosenthal, A. S.: Biologically active concanavalin A complexes suitable for light and electron microscopy. Exp. Cell Res. **70**, 443 – 447 (1972)

Sumner, J. B., Howell, S. F.: The identification of the hemagglutinin of the jack bean with concanavalin A. J. Bacteriol. **32**, 227 – 237 (1936)

Takamiya, H., Shimizu, F., Vogt, A.: A two-stage method for cross-linking antibody globulin to ferritin by glutaraldehyde. II. Antigen binding and agglutinating capacity. J. Immunol. Meth. 4, 381 – 386 (1974)

Tichá, M., Entlicher, G., Koštíř, J. V., Kocourek, J.: Studies on phytohemagglutinins. IV. Isolation and characterization of a hemagglutinin from the lentil *Lens esculenta*, moench. Biochim. Biophys. Acta **221**, 282 – 289 (1970)

Tillack, T. W., Marchesi, V. T.: Demonstration of the outer surface of freeze-etched red blood cell membranes. J. Cell Biol. **45**, 649 – 653 (1970)

Tillack, T. W., Scott, R. E., Marchesi, V. T.: The structure of erythrocyte membranes studied by freeze-etching. II. Localization of receptors for phytohemagglutinin and influenza virus to the intramembranous particles, J. Exp. Med. **135**, 1209 – 1227 (1972)

Tomita, M., Kurokawa, T., Onozaki, K., Ichiki, N., Osawa, T., Ukita, T.: Purification of galactose-binding phytoagglutinins and phytotoxin by affinity column chromatography using Sepharose. Experientia **28**, 84 – 85 (1972)

Toms, G. C., Western, A.: Phytohaemagglutinins. In: Chemotaxonomy of the Leguminosae. Harborne, J., Boulter, D., Turner, B. L. (eds.). New York: Academic Press 1971, pp. 367 – 462

Torpier, G., Montagnier, L.: Studies on the ultrastructure of concanavalin A receptor sites at the surface of normal cells and cells transformed by oncogenic viruses. Int. J. Cancer **11**, 604 – 615 (1973)

Toyoshima, S., Osawa, T., Tonomura, A.: Some properties of purified phytohemagglutinin from *Lens culinaris* seeds. Biochim. Biophys. Acta **221**, 514 – 521 (1970)

Trowbridge, I. S.: Isolation and chemical characterization of a mitogenic lectin from *Pisum sativum*. J. Biol. Chem. **249**, 6004 – 6012 (1974)

Uhlenbruck, G., Gielen, W., Pardoe, G. I.: On the specificity of lectins with a broad agglutination spectrum. V. Further investigations on the tumorcharacteristic agglutinin from wheat germ lipase. Z. Krebsforsch. **74**, 171 – 178 (1970 a)

Uhlenbruck, G., Pardoe, G. I., Bird, G. W. G.: Zur Spezifität des tumorzellcharakteristischen Agglutinins der "wheat germ lipase". Naturwissenschaften **55**, 347 (1968)

Uhlenbruck, G., Sprenger, I., Leseney, A. M., Font, J., Bourrillon, R.: Comparative studies of the anti-A agglutinins from *Dolichos biflorus* and certain snails. Vox Sang. **19**, 488 – 495 (1970 b)

Ukena, T. E., Borysenko, J. Z., Karnovsky, M. J., Berlin, R. D.: Effects of colchicine, cytochalasin B and 2-deoxyglucose on the topographical organization of surface-bound concanavalin A in normal and transformed fibroblasts. J. Cell Biol. **61**, 70 – 82 (1974)

Wada, S., Pallansch, M. J., Liener, I. E.: Chemical composition and end groups of the soybean hemagglutinin. J. Biol. Chem. **233**, 395 – 400 (1958)

Wagner, M., Wagner, B.: Immunoelektron microscopic localization of receptors with terminal N-acetyl-D-galactosamine on the surface of Ehrlich ascites tumor cells using the protection of *Helix pomatia*. Z. Immun. Forsch. **146**, 274 – 282 (1973)

Waldschmidt-Leitz, E., Keller, L.: XX. Mitteilung über Samenproteine. Zur Wirkungsweise des Hämagglutinins aus Rizinus. Hoppe Seyler's Z. Physiol. Chem. **350**, 945 – 950 (1969)

Waldschmidt-Leitz, E., Keller, L.: XXII. Mitteilung über Samenproteine. Über Toxin und Agglutinin aus Rizinus; Reinigung und Zusammensetzung. Hoppe-Seyler's Z. Physiol. Chem. **351**, 990 – 994 (1970)

Williams, M. A., Voak, D.: Studies with ferritin-labelled *Dolichos biflorus* lectin on the numbers and distribution of A sites on A1 and A2 erythrocytes, and on the nature of its specificity and enhancement by enzymes. Brit. J. Haematol. **23**, 427 – 441 (1972)

Yahara, I., Edelman, G. M.: Electron microscopic analysis of the modulation of lymphocyte receptor mobility. Exptl. Cell Res. **91**, 125 – 142 (1975)

Yariv, J., Kalb, A. J., Katchalski, E.: Isolation of an *L*-fucose binding protein from *Lotus tetragonolobus* seed. Nature (London) **215**, 890 – 891 (1967)

Young, N. M., Leon, M. A., Takahashi, T., Howard, I. K., Sage, H. J.: Studies on a phytohemagglutinin from the lentil. III. Reaction of *Lens culinaris* hemagglutinin with polysaccharides, glycoproteins, and lymphocytes. J. Biol. Chem. **246**, 1596 – 1601 (1971)

Antibody-labeling Techniques

W. D. PERKINS and J. K. KOEHLER

A. Rationale

Antibodies labeled with visual markers such as fluorochromes, ferritin, horseradish peroxidase, viruses, hemocyanin, and radioiodine, have recently been used extensively to help understand a variety of biological phenomena (RAFF, 1976; KARNOVSKY and UNANUE, 1973; WAGNER, 1973). In the field of immunology, for example, studies have focused specifically on defining lymphoid populations responsible for elicitation of both cellular and humoral immune responses. To discern these populations, antibodies were prepared against immunologic determinants unique to each cell type. Subsequently, the specific antibody was labeled with an appropriate visual marker that could be identified with the light microscope and/or transmission electron microscope (TEM). Cells responsible for cellular immune responses (thymus cells) were thus characterized as possessing surface receptors for anti-theta antibody in mice (RAFF, 1969; 1970) and cells involved in humoral immune responses (bone marrow lymphocytes) were shown to possess receptors for anti-immunoglobulin (anti-Ig) (RABELLINO and GRAY, 1971; RAFF et al., 1970; UNANUE et al., 1971). Specific characterization and localization of these and other lymphocyte receptors are currently being considered at the ultrastructural level by numerous investigators.

In addition to the immunologic studies, labeled antibodies for TEM are now used in other areas such as reproductive biology and connective tissue research. In reproductive biology, for example, questions are being asked about the arrangement of antigenic sites on sperm cell membranes and changes in these membranes during various physiologic changes (KOO et al., 1973; KOEHLER and PERKINS, 1974; KOEHLER, 1976; KOEHLER and KINSEY, 1977). In ultrastructural studies of connective tissue, labeled antibodies to procollagen have been prepared to determine where synthesis and transport of this precursor to collagen occurs (OLSEN and PROCKOP, 1974; OLSEN et al., 1973).

Although only a few examples for the ultrastructural use of labeled antibodies have been presented, it is clear that the use of labeled antibodies to attack biological problems can be a useful approach where there is a need to identify a highly specific molecular species with a level of resolution beyond the limit of the light microscope.

B. Antibody Labels

For any research project that requires the use of labeled antibodies, it is important to carefully evaluate the label to be used since each has its advantages and disadvantages. One of the first electron-opaque markers to be used for labeling antibodies was ferritin (SINGER, 1959). Ferritin has a molecular weight of about 450,000, a diameter of approximately 120 Å, and contains an electron-opaque iron core. Since its appearance as an antibody label, it has become the most commonly used label available for ultrastructural studies and has been extensively utilized in membrane studies (DE PETRIS and RAFF, 1973; KARNOVSKY et al., 1972). In thin section or 'whole mount' preparations, ferritin molecules appear as individual discrete round opacities. Because each molecule can be defined, the apparent location of antibody receptors can be determined and, in fact, most investigations routinely equate the location of a ferritin molecule with an antibody receptor site. It is, therefore, important to be aware of the limitations that are shared by this molecule and all other large antibody markers. It is possible that steric effects may alter the distribution of molecules. Second, during conjugation procedures, more than one antibody molecule may be complexed with one marker molecule and vice versa. Third, if viable cell membranes are labeled, relocation of receptor sites may occur. Even with these limitations, however, much valuable information has been obtained using large molecules as antibody markers.

Another commonly used antibody label is horseradish peroxidase (HRP). Its use as a marker is based on the formation of an electron-opaque reaction product from oxidation of a substrate (3-3′ diaminobenzidine) in the presence of hydrogen peroxide (GRAHAM and KARNOVSKY, 1966). Although HRP has a low molecular weight (40,000), point-to-point discrimination of specific antibody receptor sites cannot be defined. The lack of specificity is a result of formation of an indeterminate amount of reaction product. If precise site specificity of antibody receptors is not a criterion, then HRP can be a useful antibody label (REYES et al., 1975; MATTER et al., 1972; AVERAMEAS and TERNYNCK, 1971; AVERAMEAS, 1969).

Viruses such as tobacco mosaic and the phages have also been used as antibody markers with the TEM and scanning electron microscope (SEM) because of their unique shapes. They are, perhaps, less useful because of their large size (KAY, 1975; NEMANIC et al., 1975; HAMMERLING et al., 1975).

Hemocyanin is an excellent antibody label because of its cylindrical shape. It is also advantageous in that it can be localized to about the same degree as ferritin (SMITH and REVEL, 1972; KARNOVSKY et al., 1972; UKENA et al., 1974; KOEHLER and PERKINS, 1974; KOEHLER, 1976).

Radioiodine is another antibody marker for TEM. Its presence can be detected after radioautographic procedures as electron-opaque silver grains. One distinct advantage that radioiodine has over most other TEM antibody labels is that it can also be used for light-microscopic as well as TEM study. Ultrastructurally, it has a lower resolution than some of the other markers.

C. Methods for Coupling Label to Antibody

Once an appropriate marker has been selected for antibody coupling, a method for conjugation must be determined. The technology for labeling antibodies for TEM is presently being developed at a rapid rate and is, therefore, undergoing constant change. However, two basic methodologies have been advanced that form the basis of most labeling procedures. These include a one-step and a two-step method.

I. One-Step Method

In the one-step method, antibody and label are simultaneously mixed with a coupling agent (e. g., glutaraldehyde) (AVERAMEAS, 1969; TAKAMIYA et al., 1974). In this procedure the possibility exists for formation of antibody-label, antibody-antibody, and label-label conjugates. Thus, unwanted conjugates could modify experimental results. However, further purification of the wanted antibody-label conjugate may be realized by chromatography, ultracentrifugation, and immunoelectrophoresis (WAGNER, 1973).

II. Two-Step Method

In the two-step method the label is first treated with a coupler and then antibody is reacted with label-coupler complex (OLSEN et al., 1973; AVERAMEAS and TERNYNCK, 1971; KISHIDA et al., 1975; OTTO et al., 1973; TAKAMIYA et al., 1974). In this manner, antibody-antibody and label-label conjugates can, for the most part, be avoided. The particular biological problem to be solved may dictate which method should be used. If point-to-point localization is not a consideration, the one-step procedure may then be the method of choice because it requires fewer preparative steps. It has been found, for instance, that for identification of lymphocyte populations the one-step method is sufficient (UUNANUE et al., 1972 b). However, if receptor site mapping of single cells is being considered, the two-step method may be more appropriate.

General procedures for labeling with ferritin and HRP are dealt with in another Chapter of this volume (NICHOLSON, 1977). Therefore, it is the purpose of this discussion to present methodologies for labeling antibodies with the markers hemocyanin and radioiodine for TEM examination. The labeling techniques to be considered are those most commonly used at the present time. Their advantages and disadvantages will be evaluated. In addition to antibody labeling procedures, the further use and processing of several kinds of labeled antibody will be discussed. The procedures discussed will include preparation of cell suspensions prior to reaction with labeled antibody, reaction of cells with labeled antibody, and specialized preparation of cells after reaction with labeled antibodies.

D. Iodination of Antibody

Recently, radioactive iodine (^{125}I) has been used as an ultrastructural label for antibody. It is especially useful for morphologic studies in that it can be used for both light-microscopic and ultrastructural evaluation. This has the advantage of allowing one to observe results almost immediately with the light microscope including the quantitation of large numbers of cells. At the ultrastructural level, ^{125}I label has the advantage over other markers in that the label (silver grains) can be observed on many cells very quickly at a low magnification ($\times$ 2000). In contrast, nonradioactive markers such as ferritin require that each cell be carefully scrutinized at relatively high magnification ($10-15\times10^3\ \times$), requiring many tedious hours. Another advantage of radiolabeled antibody is that scintillation counting can be carried out on antibody-labeled cell populations to determine the total antibody uptake. In the past, most ultrastructural radioautographic procedures had the serious disadvantage of requiring months for exposure to nuclear emulsion to produce enough silver grains to be seen in the TEM. However, with the procedures now available for labeling antibody with ^{125}I (GREENWOOD et al., 1963; PERKINS et al., 1972), the antibody can be labeled at a high specific activity, which can reduce exposure time to a few days. Perhaps the only real disadvantage of using a radioactive label is that its resolution is about 1000 to 2000 Å (CARO and VAN TUBERGEN, 1962). This lower resolution results in part because the nuclear emulsion over the cell containing silver halide crystals can be reduced to metallic silver at some distance from the radiation emitting source. Thus, a silver grain does not necessarily represent the exact point where radiolabeled antibody is located. The three most common methods for labeling antibody with ^{125}I are a modified chloramine T procedure, labeling with lactoperoxidase, and a conjugation method. Each procedure has advantages and disadvantages that are discussed (videinfra). As the chloramine T procedure is utilized routinely in this laboratory, this method will be described in considerable detail.

I. Iodination of Antibody with Chloramine T

Because iodine is concentrated in the thyroid gland and radiation damage could occur if ^{125}I were accidentally inhaled or spilled on the body surface, extensive safety precautions should be taken during this procedure. The procedure should be carried out using a monitor (sodium iodide Geiger counter) in a hood and utensils should be disposable.

The chloramine T (CT) method of labeling antibody for ultrastructural analysis is a modification of the Greenwood procedure (PERKINS et al., 1972). In this procedure ^{125}I is incorporated into antibody by direct substitution into tyrosyl residues of the protein. This is accomplished by oxidation with CT and reduction with sodium metabisulphite.

The antibody to be iodinated is placed in a 5-ml disposable plastic tube and in most instances, the amount of protein used is not greater than 300 μg and is contained in less than 100 μl of phosphate-buffered saline (PBS). For ease in handling, carrier-free ^{125}I (New England Nuclear, Boston, MA) is obtained in a 'V' vial in less than 100 μl of solution. For the iodination reaction, the iodine must be used at a neutral pH to prevent denaturation of the protein, therefore the stock solution must be neutralized since it is supplied in alkaline solution. The stock bottle is neutralized with 1 N NaH_2PO_4 and approximately 5 μl of 1 N NaH_2PO_4 will neutralize 30 μl of $Na^{125}I$. After neutralization the iodine is added to the antibody.

II. ^{125}I-Labeled Antibody for Transmission Electron Microscopy

For labeling of cells for TEM, approximately 1 mCi of $Na^{125}I$ is added for every 20 μg of antibody. This will usually provide enough specific activity (15 – 20 μCi/μg antibody) to allow for an exposure time of less than one week and less than one day for light microscopy. Chloramine T, 50 μg (10 μl of a 10 mg/2 ml PBS solution), is added to the antibody and $Na^{125}I$, which initiates the reaction. Since the reaction involves release of volatile iodine the reaction tube should be covered immediately and the hood window closed during this reaction. The reaction is carried out at 4° C for 15 min. It is then stopped with 100 μg (2 ml of a 10 mg sodium metabisulfite/2 ml PBS solution) of sodium metabisulfite. One drop of a 1% cold iodine in PBS solution is added to dilute any residual reaction. The labeled antibody is now removed from unreacted $Na^{125}I$ on a G-25 Sephadex column packed in a 5-ml disposable plastic pipette equilibrated with PBS. To know when the antibody comes off the column, a visual marker of 1 drop of hemolyzed serum that is known not to react with the antibody can be added to the reaction mixture. Alternatively, a sodium iodide Geiger counter can be used to monitor the movement of the antibody as it passes through the column. In the majority of experiments most of the antibody is recovered in less than one ml of volume. This

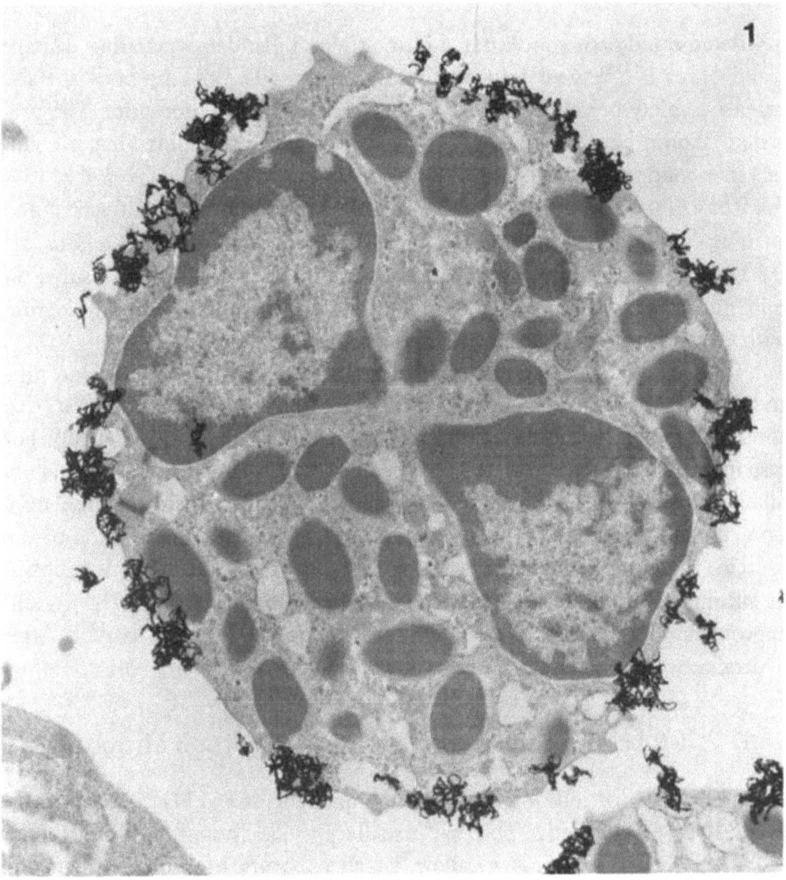

Fig. 1. A radioautograph of a bone marrow leukocyte that was labeled with ^{125}I-labeled anti-Ig at 4° C (specific activity 15 μCi/μg of antibody). Radioautographic exposure time was less than one week. Note numerous silver grains that represent the labeled antibody overlying the plasma membrane. ×10,350

can usually be used directly without further concentration. The specific activity of the labeled antibody can be determined by making a 1 : 1000 dilution (10 μl labeled antibody/10 ml PBS) and counting it in a scintillation counter. A resulting specific activity of 15 to 20 μCi/mg of antibody is ideal for short periods of exposure when doing radioautography for TEM. The total procedure can be accomplished from beginning to end in less than 1½ h. Possible drawbacks to this method may be a partial loss of antibody activity due to denaturation or modification of reactive components of the antibody molecule. Denaturation may occur because of protein oxidation and reduction during the reaction or by direct contact with the neutralized $Na^{125}I$ solution.

It is also possible that tyrosine may be part of the reactive portion of the antibody molecule and iodination could therefore alter the reactive site in such a way that antigen-antibody complexing could not take place. Experience has shown, however, that for ^{125}I label of anti-immunoglobulin antibody, activity is still very adequate for most studies (Fig. 1).

III. Lactoperoxidase Labeling of Antibody

Another method for labeling with ^{125}I is based on enzyme-catalyzed halogenation. In this procedure lactoperoxidase is used for iodination of antibody in the presence of hydrogen peroxide (MARCHALONIS, 1969). This method has been purported to allow for ^{125}I label of high specific activity and less denaturation of protein. However, the antibody to be labeled is subject to the same direct contact of potentially noxious ^{125}I solution that is used in the CT method.

IV. Antibody Labeling with an Acylating Agent

To obviate the possibility of denaturation of protein during iodination by direct contact with the ^{125}I solution and oxidation, a method has been developed in which ^{125}I is first incorporated into an acylating agent, followed by reacting the ^{125}I-labeled acylating agent with the protein to be labeled (BOLTON and HUNTER, 1973). In the latter reaction, free amino groups in the protein molecule link ^{125}I-labeled groups by amide bonds. Although this procedure involves many steps, it does prevent for the most part denaturation of the protein. It does, however, have the same problem involved in the CT and lactoperoxidase procedures, in that reactive sites on the antibody molecule could be modified with loss of activity due to incorporation of ^{125}I-labeled groups into these areas.

The three methods for ^{125}I labeling of antibodies described above are those most commonly used. Each procedure has its advantages and disadvantages. To specifically indicate that one is better than another is inappropriate since for a particular antibody one method may give a better result than another. Details were provided for ^{125}I iodination using the CT method because it is simple, reproducible, and for most antibodies it gives satisfactory results. This method might, therefore, be tried before more complex procedures are considered.

E. Hemocyanin Label of Antibody

Another antibody label that is used for ultrastructural study is the electron-opaque molecule hemocyanin. It is the oxygen-carrying unit of marine whelks that contains copper and has a unique cylindrical structural configuration (VAN BRUGGEN et al., 1963). This configuration allows for easy identification at the ultrastructural level. Because it can be readily identified it is especially useful in studying the orientation and movement of cell surface antigens when used in conjunction with surface replica techniques (videinfra).

Hemocyanin is obtained from the marine whelk *Busycon canaliculatum* (Woods Hole Biological Supply, Woods Hole, MA). This whelk is preferred over some other species because its hemocyanin molecule is large, relatively stable, and is easily discernible at the TEM level. Its molecular weight is approximately 9×10^6 daltons and it is composed of subunits of about 280,000 daltons. Its appearance is either cylindrical or rectangular depending on whether it is viewed along its axis or from the side. It has a diameter of about 300 Å (VAN BRUGGEN et al., 1963).

I. Purification of Hemocyanin

The hemocyanin is collected from the marine whelks by incision of the foot and draining the hemocoel fluid (KARNOVSKY et al., 1972). One dozen *Busycon canaliculatum* yield about 300 cc of hemocoel fluid, which provides, after purification, about 2500 mg of hemocyanin. During the initial collection of hemocoel fluid, coarse particulates accumulate. This macroscopic debris is removed by passing the fluid through a vacuum filter using coarse filter paper or several thicknesses of gauze. Smaller debris is removed by centrifugation at 3000 rpm (IEC P. R. 6000 refrigerated centrifuge at +4° C). The resulting supernatant is centrifuged at 10,000 rpm for 20 min (Spinco Ultracentrifuge, Beckman Instruments Inc., Spinco Div., Palo Alto, CA). At this stage, a fine white precipitate is removed from the blue hemocyanin-containing supernatant. The resulting supernatant is centrifuged at 100,000 *g* for 1 h and the supernatant discarded. The remaining pellet of hemocyanin is 'jelly-like' and does not dissolve easily. Therefore, the pellet is dissolved by standing overnight (4° C) in 3% NaCl solution. Alternatively centrifugation at slower speeds can be used to concentrate the hemocyanin without pelleting. The hemocyanin molecule, which is made up of smaller subunits, is rather unstable at low ionic strength or at extremes of pH and tends to deaggregate. A 3% NaCl solution at pH 6.9 to 7.0 appears to stabilize the molecule for long periods of time.

The hemocyanin in 3% NaCl is finally passed through a Millipore filter (0.45 μ) for sterilization. At this time the hemocyanin can be stored at +4° C until needed for conjugation. In this condition, the hemocyanin is usable for at least 6 months. One need not be concerned that the deep blue stored solution may turn white due to the deoxygenation of the hemocyanin.

The blue color will return if the solution is gently swirled in contact with air. A low concentration of sodium azide may be added to discourage growth of microorganisms.

II. Conjugation of Hemocyanin with Antibody

The conjugation of hemocyanin to antibody is based primarily on a procedure developed by KARNOVSKY et al. (1972). The hemocyanin salt solution is dialyzed with 0.1 M PBS, pH 6.9. After dialysis, the hemocyanin-containing solution that has been corrected for volume is mixed with antibody in a ratio of 1 part antibody : 9 parts hemocyanin. The volume of the mixed antibody and hemocyanin is brought to 50 ml in 0.1 M PBS pH 6.75. To this solution 1 ml of 5% glutaraldehyde is slowly added dropwise (1 ml over a 5-min time period). The resulting 0.1% glutaraldehyde 'conjugate' solution is stirred at room temperature over a 2-h period. The solution is dialyzed against 0.1 M PBS overnight and is then centrifuged at 2000 rpm to remove any precipitate that may have appeared. The final supernatant is then filtered through a Millipore filter, pore size 0.45 μ, to sterilize the conjugated antibody-hemocyanin complex. The conjugate can now be stored at +4° C. It is important not to freeze the material because the hemocyanin will denature. The resulting hemocyanin antibody complex can now be reacted with the appropriate tissue. After preparation of the cells for ultrastructural analysis, the cylindrical or square shape of the hemocyanin molecule, presumably attached to the antibody, can clearly be observed (Fig. 2).

F. Reaction of Antibody with Cells

Labeled antibody can be reacted with viable cells or cells fixed with low concentration of glutaraldehyde. Fixed cells have the obvious disadvantage of receptor site loss due to denaturation by the fixative. It has been calculated that 40 to 50% of receptor sites may be lost due to fixation (KRAEHENBUHL and JAMIESON, 1972). In either case, the cells are washed 3 times in buffer (Hank's balanced salt or similar solutions). Centrifugation is routinely used during the washing of viable cells at 1000 rpm for 6 min in a clinical centrifuge. Once the cells have been washed they are reacted with labeled antibody. The amount of labeled antibody to be reacted for a given number of cells will vary for the kind of antibody and its specificity for a given antigen. For example, it has been determined that reacting bone marrow-derived lymphocytes with labeled anti-Ig antibody requires 10 μg of antibody for 10^7 cells (PERKINS et al., 1972). The reaction is carried out in a volume of less than 100 μl. Once the optimum amount of antibody is determined, the cells are reacted

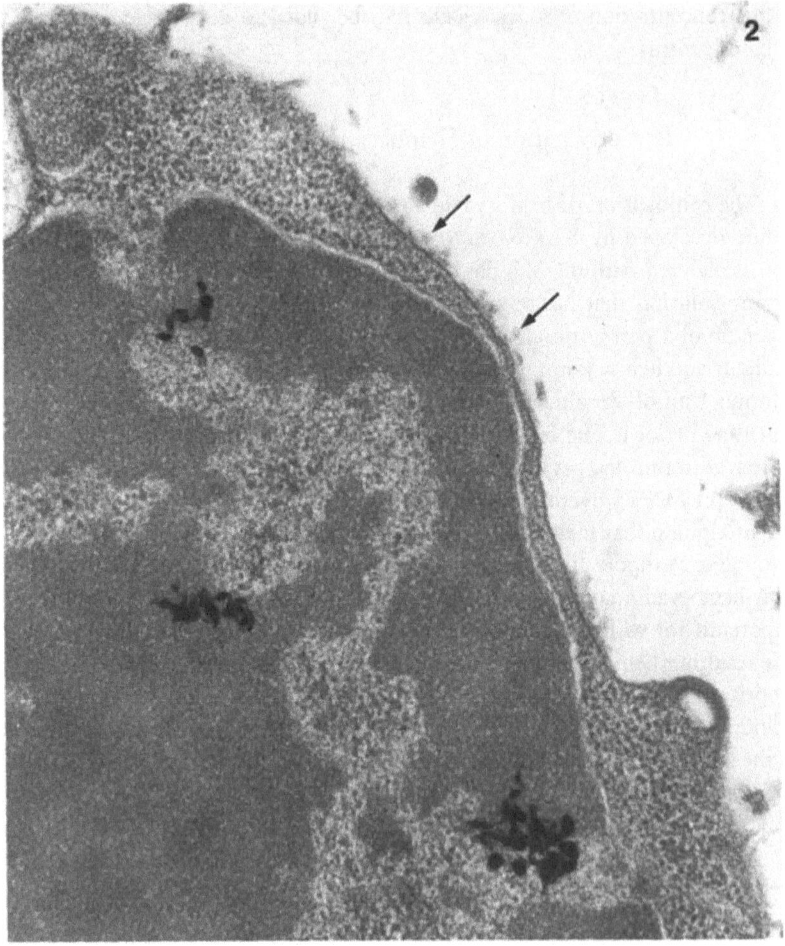

Fig. 2. A radioautograph of a bone marrow lymphocyte labeled sequentially with two markers. The cell surface is labeled with hemocyanin conjugated to anti-Ig. Note unique square shape of molecules on cell surface (*arrows*). The silver grains over the nucleus respresent ^{3}H-uridine. ×63,000

for 30 min at 4° C. After 30 min the cells are washed 3 or 4 times to remove uncomplexed antibody. At this point, the cells can be fixed and prepared for TEM. If receptor site movement or rearrangement is being studied, the reacted cell suspension can be warmed to 37° C for a period of time before the fixation step (UNANUE et al., 1972 a, b and KARNOVSKY et al., 1975) (Fig. 3). Unlike ferritin, which has sufficient inherent contrast to be visualized in unstained sections, hemocyanin requires special processing to make it

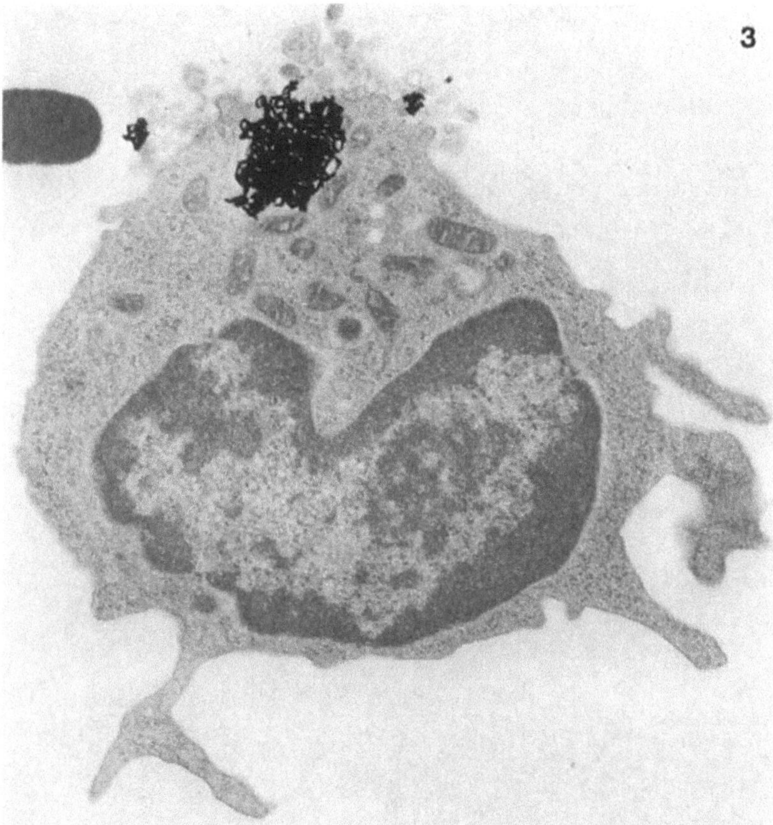

Fig. 3. A radioautograph of a lymphocyte reacted with ^{125}I-labeled anti-Ig and warmed to 37° C for 15 min prior to fixation. Note endocytosis of ^{125}I-labeled anti-Ig and compare arrangement of silver grains with Figure 1, which was reacted at 4° C. ×6,300

visible in the TEM. One such method involves staining with a half saturated aqueous uranyl acetate solution for 2 to 3 h at 60° C. Obviously, precautions must be used to prevent drying of the preparation such as enclosing the specimen in sealed, humidified containers.

G. Clotting Procedure for Handling Single Cell Suspensions

Preparation of single cell suspensions for TEM is a time-consuming and cumbersome procedure involving many centrifugations. Cell loss and damage

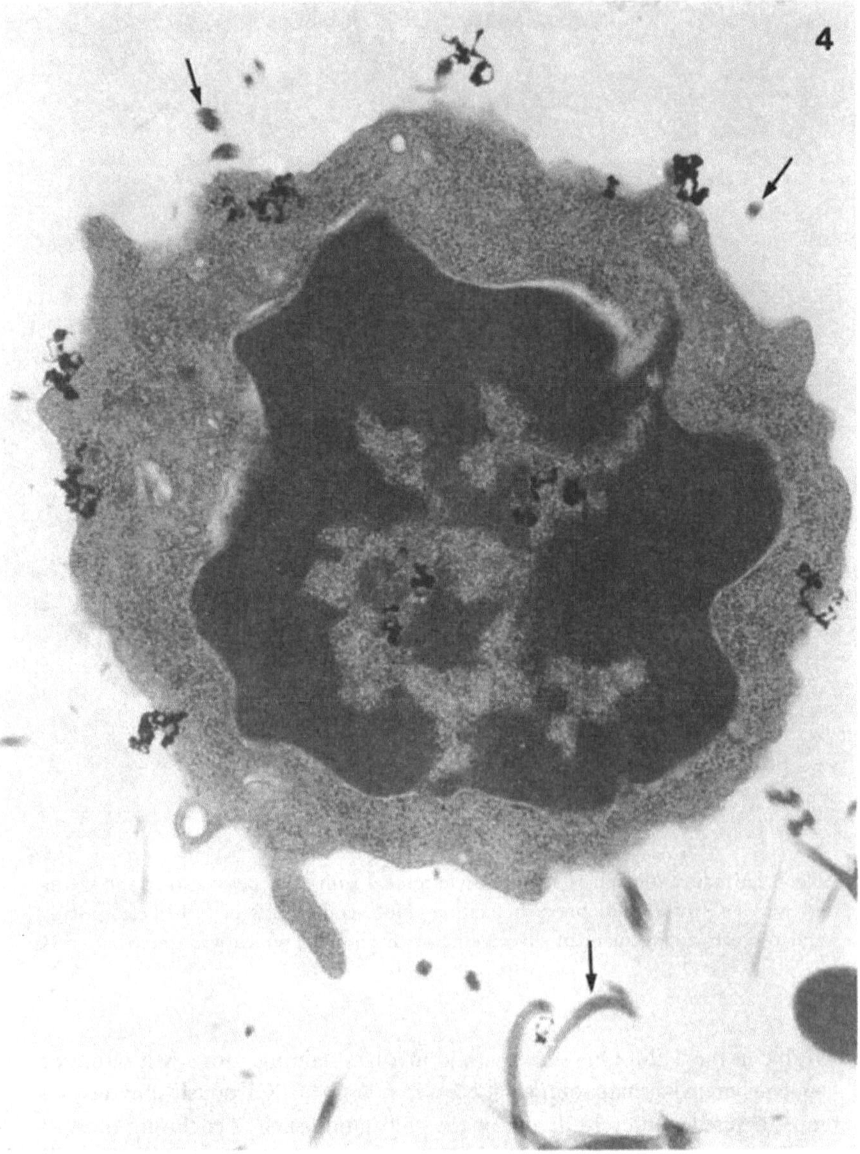

Fig. 4. A radioautograph of a short-lived bone marrow lymphocyte that was labeled with ^{125}I-labeled anti-Ig as a cell surface marker and also ^{3}H-thymidine as a nuclear label to measure life span. The cell was processed for TEM by embedment in a fibrin clot. The strands of fibrin (*arrows*) do not interfere or detract from the information provided. ×18,000

is not uncommon. To circumvent these problems, a clotting procedure has been developed that allows the cells to be handled as blocks of tissue during processing for TEM.

The principle of this procedure is to produce a fibrin clot that will entrap cells. Once the cells are enmeshed in the fibrin, the fibrin and cells can be fixed and processed as blocks of tissue. This procedure is used at the time of initial fixation of antibody-reacted cells. The cells are fixed for a short period of time (15 min) in glutaraldehyde-formaldehyde (KARNOVSKY, 1965). The cells are then washed 3 times in 0.1 M cacodylate buffer. After the last centrifugation the supernatant is poured off and the remaining cell pellet is resuspended in the residual volume on a vortex. The cells are then embedded in a clot. The stock solutions for this procedure include:

- NaCl buffer pH 7.35 – 0.15 M NaCl
  0.02 M imidazole
- Fibrinogen stock (Sigma Chemical Corp.) – 2 mg fibrinogen
  1.0 ml 0.15 M NaCl buffer
- Thrombin stock (Park-Davis) – 0.5 μg thrombin
  1.0 ml 0.15 M NaCl buffer

The fibrinogen and thrombin may be prepared and stored frozen at –70° C or prepared just prior to use. The following dilutions and volumes of fibrinogen and fibrin were established for 10^7 cells and can be varied relative to cell number. The stock fibrinogen solution is diluted 1 : 4 with 0.15 M NaCl buffer; 0.2 cc of this dilution is added to the fixed cell suspension. Subsequently, 0.1 cc of thrombin stock is added to the cells. The cells, fibrinogen, and thrombin are immediately vortexed. This initiates fibrin formation. The fibrin clot is allowed to form at room temperature for 10 min. Once the clot has formed, it is placed in fixative, formaldehyde : glutaraldehyde (1 : 1) on a depression slide (Maximow) for 15 min. After this time period the soft jelly-like clot will be solid enough to cut into 1-mm blocks. These blocks may be further processed for TEM, radioautography, or other conventional methods. When cells are prepared for TEM in this fashion, the background strands of fibrin should in no way interfere with observations in the electron microscope (Fig. 4). If the fibrin strands do interfere (e. g., studies related to cell surfaces), the clot density can be reduced by changing the dilution of stock fibrinogen (e. g., 1 : 10 vs. 1 : 4). Experience has shown that it is convenient to produce a few 'trial' clots with different dilutions just prior to an experiment. In this manner, the clot density can be determined prior to treating the cells.

H. Radioautography

After conventional embedment procedures, tissues that require radioautography, e. g., cells reacted with ^{125}I-labeled antibody, can be processed in

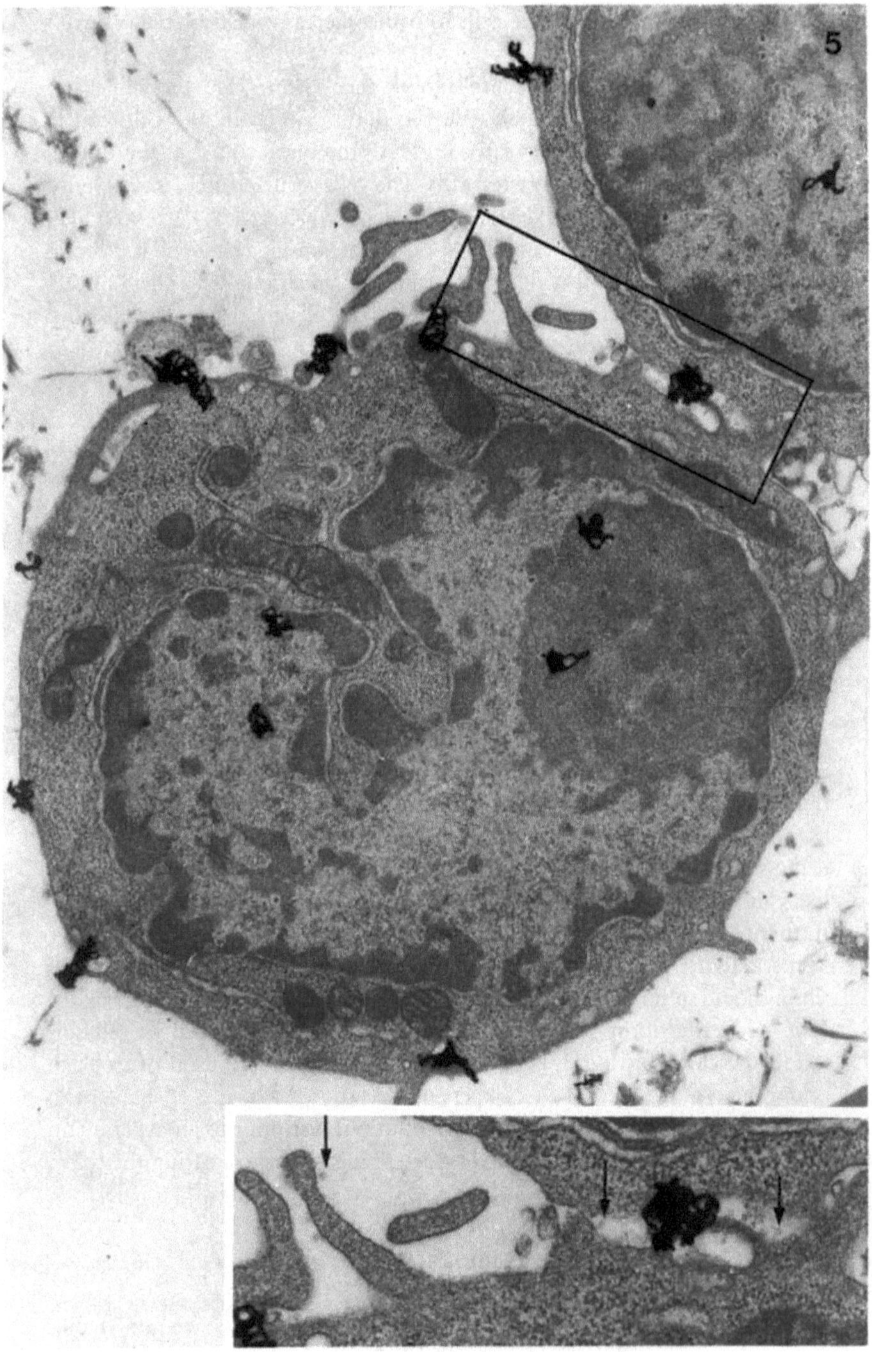
5

the following manner. Thin sections (silver gold–gold) are cut and placed on *bare* copper grids. The grids are coated with Ilford L-4 emulsion (Ilford, Ltd., Ilford, Essex, England), diluted 1 : 1 with distilled water. The grids can be coated with a platinum wire loop that contains a thin film of wet emulsion. The coated grids are stored from 5 to 10 days at 18° C (CARO and VAN TUBERGEN, 1962). After this exposure time, the grids are developed with Kodak K-19 developer for 3 min, washed briefly in distilled water, fixed in full strength Kodak acid fixer for 1 to 2 min, and again washed in distilled water. Once development is completed, it is important that the grids be kept wet until after staining, otherwise the emulsion over the sections can be seen and may totally obscure the tissue to be studied. Therefore, the grids are usually taken directly from the last wash of the development procedure and are floated on or immersed in distilled water contained in small petri dishes. The wet grids are immediately stained in Reynolds' lead citrate (REYNOLDS, 1963). After staining, the sections are ready for study in the TEM (Fig. 5). Standard works on the details of radioautographic technique should be consulted before attempting these procedures (SALPETER and MCHENRY, 1973; BUDD, 1971).

I. Replica Techniques

The procedure of making a cast of a specimen with inert materials and its examination in the electron microscope has been a basic fine structural preparative method for several decades. MAHL (1940) was among the first to employ this technique for metallurgical samples, and WILLIAMS and WYCKOFF (1946) developed the complimentary shadowing method and applied it to biological material. This technique involves the preparation of a thin film bearing an accurate reproduction of the original sample and suitable for direct observation in the electron microscope. These procedures have in common the use of a vacuum deposition apparatus in order to shadow the specimen, thus introducing contrast into the replica, and the evaporation of a carbon 'backing' layer to stabilize the replica.

Although the replica method was among the earliest preparative procedures used in biological electron microscopy, it was largely eclipsed by the development of thin sectioning. The advent of methods for coupling topologically detectable macromolecules to immunoglobulins (SINGER, 1959) has,

Fig. 5. A radioautograph of a cell containing nuclear label (^{3}H-thymidine as an indicator of DNA synthesis) and possessing two cell surface markers. One surface marker is ^{125}I-labeled anti-Ig and is represented by silver grains. The other surface marker is hemocyanin-labeled anti-theta antibody. ×13,500. Note rectangular shape of hemocyanin (inset: *arrows*). ×39,500

however, made the replica approach very useful for examining the localization of antigenic sites on cell surfaces. The prerequisites for successful localization of surface-bound markers in replicas include physiochemical considerations common to all labeling studies such as the stability of the complex with respect to pH, temperature, ionic strength, and stearic considerations relative to binding of the complex to specific sites. Of particular importance to the replica technique are morphologic parameters pertaining to the ease of recognition of the marker in replicas. Thus, the size and shape of the labeling molecules must be carefully considered with regard to recognition phenomena. The degree of resolution required of the particular sites of interest must also be considered together with such recognition factors. It would obviously be simpler to detect a very large marker complex than a small one, but the site separations one wishes to resolve may not be labeled with structures having larger dimensions than nearest neighbor sites. Similarly, highly asymmetric structures such as rod-shaped viruses (KOO et al., 1973) have been utilized as labels and are easily recognized but have a relatively limited degree of inherent resolution. One must choose the marker that will best serve the purpose of the particular investigation and will provide the optimal recognition possibilities consistent with desired resolution. Two of the most popular marker molecules used in conjunction with replica procedures are ferritin and hemocyanin. Ferritin has been used extensively since the pioneering work of SINGER (1959) in immunolabeling studies but has only recently been employed in replica work, particularly in conjunction with the freeze-etching method (see below). Hemocyanin was initially used by SMITH and REVEL (1972) in marking lectin-binding sites and has been used extensively by KARNOVSKY et al. (1972) to identify immunoglobulin sites in lymphoid cells in standard as well as freeze-etching replicas. See earlier sections for details regarding the labeling of immunoglobulins with ferritin or hemocyanin.

I. Surface Replica Technique

The simplest procedure for examining surface markers on cell surfaces involves the fixation of the cell population after the labeling regimen in one of the standard glutaraldehyde mixtures followed by washing several times in distilled water. The cells at an appropriate dilution are then deposited in a droplet on a parlodian-coated electron microscope grid held by forceps (see Fig. 6). Most of the fluid is blotted away leaving a very thin film of suspension, which is allowed to dry. The grid is then carefully attached to the edge of a strip of double-sticky Scotch tape fastened to a coverslip or a broken half of a glass microscope slide. The slide is placed on the stage of a vacuum evaporator and a replica is prepared according to conventional procedures. A detailed discussion of replication hardware and methodology can be found in a number of standard works (BRADLEY, 1965; HENDERSON and GRIFFITHS,

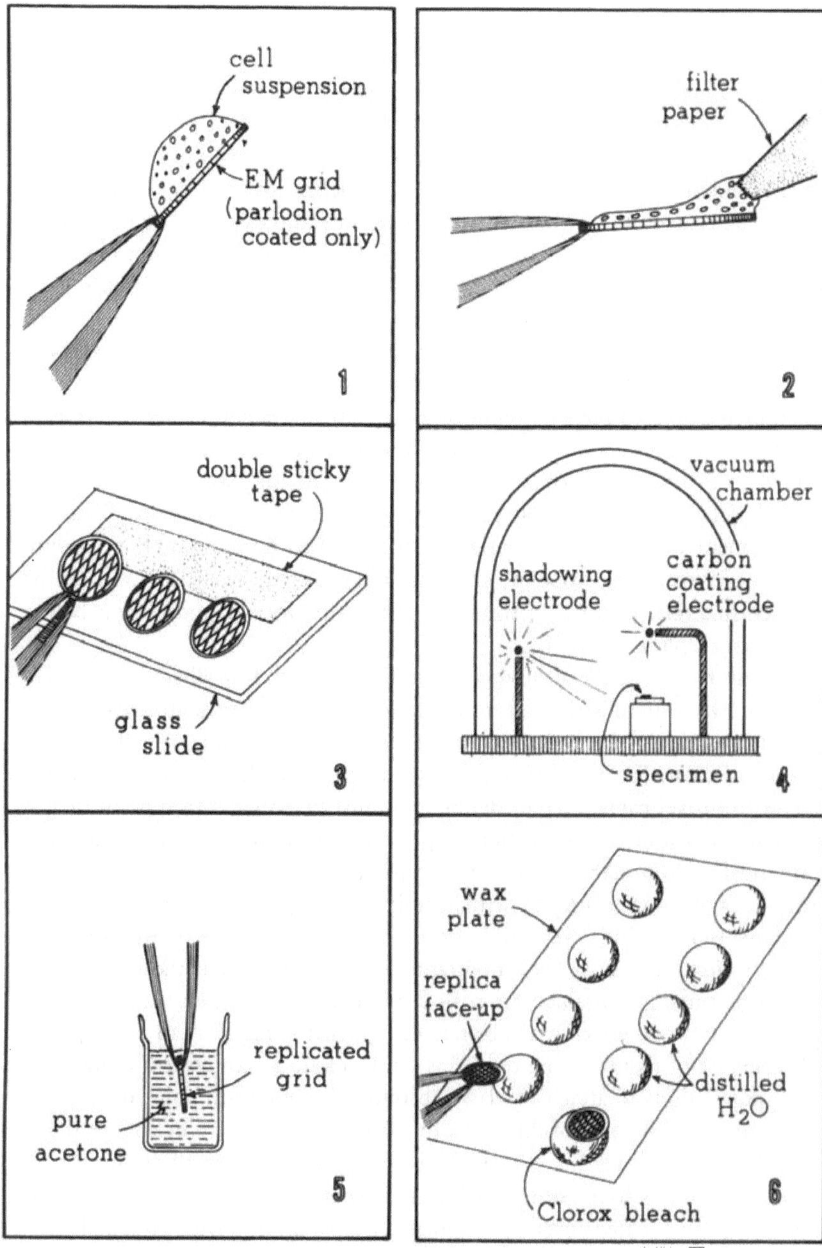

Fig. 6. Schema showing a method for preparing replicas for TEM. The cells are deposited in a dilute drop onto a parlodion-coated grid (*1*); blotted almost to dryness with a bit of filter paper (*2*); attached to a carrier after drying (*3*) and replicated in vacuo (*4*). The finished replica is then dipped gently into acetone for 10 to 15 s to remove parlodion (*5*) and washed with clear water (*6*). See text for further details

1972) and will not be attempted here. It should be mentioned, however, that the carbon-platinum procedure is probably the method of choice in producing finer grain sizes than can be achieved with other metallic shadowing materials. If electron-beam shadowing systems are available, they have been described as producing even better results (ABERMANN et al., 1972) and are faster and more reproducible. In addition to the shadowing step, the specimen must be coated uniformly with a thin layer of carbon to stabilize the replica. This is deposited by a second electrode pair positioned directly over the specimen in much the same way that standard carbon-coated grids are prepared.

After the grids are removed from the evaporator, they are carefully picked up and gently dipped into acetone for 10 s to remove the parlodion, allowed to dry, and then floated (replicated surface up) on a drop of Clorox or other commercial bleach preparation in order to remove cellular debris. This cleaning process takes about ½ h and is followed by flotation of the grids serially on 6 to 8 drops of distilled water to remove all the bleach. The grids are dried and can be immediately examined in the electron microscope. Thus, an experiment done in the morning can be evaluated by this simple procedure in the electron microscope the same afternoon. Alternatively, the cells can be stored in a refrigerator after fixation and washing for a period of time before preparing replicas.

Unfortunately, the air-drying technique is useful only when dealing with certain types of cells that are resistant to shrinkage. Spermatozoa of many species withstand this treatment rather well, and an example of such a preparation is shown in Fig. 7. Most other somatic cells, including blood elements, tend to shrivel badly after air drying and must be either critical point or frozen dried before the replication step. Since the critical point method has emerged as a routine method for the preparation of SEM specimens, many laboratories are now equipped with simple critical point instruments so that the inclusion of this step should not seriously effect the ease of preparing surface replicas. It is, however, important to bear in mind that the labeled cells must be suitably contained for critical point drying in a way that will allow them to be transferred to a substrate on which the replicas can be prepared. In the case of tissue culture cells, this can be easily accomplished by allowing cell populations to attach to glass substrates before critical point drying. Alternatively, Millipore (Millipore Corp., Bedford, MA) or Nucleopore filters (Arthur H. Thomas Co., Philadelphia, PA) as well as small encapsulating mesh containers have been used for carriers. It is also possible to affix cells onto a gelatin-coated coverslip fragment by cross-linking with glutaraldehyde (KINSEY, pers. commun.) or to utilize polylysine-coated substrates for cell attachment. Most simply, the cell can be carried through the critical point procedure in a small lens tissue bag and sprinkled onto parlodion-coated grids after drying. The grids can then be processed in the same manner as described for air-dryed samples.

Fig. 7. A rabbit spermatozoon labeled with an antibody-hemocyanin complex and prepared by the surface replica method. The distinctive hemocyanin markers can be seen heavily distributed over the sperm head surface. The inset clearly shows the 'square' shape of the hemocyanin molecules and a characteristic central dimple or groove. ×18,900. Inset, ×63,000

II. Freeze-etching Technique

The freeze-fracture and -etching procedure has taken its place as an important ultrastructural method, particularly in the area of membrane research. Details of freeze-etch technology and procedures have been covered in a number of reviews (BULLIVANT, 1973; KOEHLER, 1972) and are beyond the scope of this Chapter. The interpretation of freeze-fracture images has been greatly aided by the now widely accepted 'membrane splitting hypothesis' (BRANTON, 1966), which suggests that internal membrane architecture is revealed during the fracturing of low-temperature specimens. Due to this phenomenon, the true surfaces (either internal or external) of membranes are not exposed by freeze-fracture and an additional 'etching' step must be used to reveal surface-bound markers such as ferritin or hemocyanin-labeled antibodies (Fig. 8).

Thus, after the fracturing step, the ice surrounding the cell in question is partially removed by sublimation in vacuo, thereby exposing a portion of the true surface. If that region contains marker molecules of a characteristic size and shape, they can be easily identified by this technique. In addition to simply seeing the surface distribution in freeze-etched preparations, it is sometimes possible to make correlations between the loci of intramembranous components such as intercalated particles and the surface sites (see, e. g., PINTO DE SILVA et al., 1971).

In contrast to the standard surface replicas, an obvious disadvantage of this technique is that only small fragments of a given cellular membrane can be examined at one time, thus making it difficult to determine overall patterns without extensive statistical sampling methods. This is particularly difficult in the case of highly asymmetric or irregular cells.

The advent of higher resolution SEMs also opens the possibility of seeing surface markers directly in secondary electron mode without the need of going through the extra steps required in replica production for TEM. Indeed, in some instances, it would be difficult or impossible to prepare surface replicas whereas SEM would be straightforward. Such is the case, for example, in examining the distribution of surface label on cells within an organ or tissue not presenting a surface. This methodology is discussed in detail elsewhere in this volume (BROWN and REVEL, pp. 65 – 88).

J. Conclusions

The techniques that have been presented are now developed to such a degree that valuable, reproducible biological data can be obtained. In this regard, it is now practicable by labeling techniques to monitor more than one cell parameter at the same time. For instance, RNA synthesis can be followed

Fig. 8. Freeze-etching preparation of a rabbit spermatozoon labeled with antibody-hemocyanin complex. The initial fracture ridge can be seen at the arrow with the exposed A face of the plasma membrane extending to the right. Deep etching (3 min, –95° C) has revealed the true surface of the cell to the left of the fracture ridge. The hemocyanin molecules can be easily recognized on the surface. ×58,500

and the cell involved in this synthesis can be defined (Fig. 2). Also, the life span of a specific kind of lymphocyte can be determined (Fig. 4). It is even possible to utilize more than two markers of cell activity. For example, Figure 5 illustrates a cell whose life span was determined by ^{3}H-thymidine incorporation and two cell-surface markers used for specific cell identification.

Perhaps the one major hurdle that remains to be solved is the use of labeled antibodies for TEM intracellular studies. The problem arises because macromolecular antibody-label complexes do not easily penetrate cell membranes. It is difficult, therefore, unless the cells are viable and can endocytose the macromolecular conjugates, to get labeled antibody into cells. It is clear, of course, that endocytosis studies would only be appropriate for certain categories of investigations. Although attempts have been made to develop techniques for intracellular use of labeled antibodies, the procedures are complex and, for the most part, the results are not entirely satisfactory (PAINTER et al., 1973; KRAEHENBUHL and JAMIESON, 1972). One of the more promising methods is related to studies on collagen synthesis (OLSEN and PROCKOP, 1974; OLSEN et al., 1975).

Because of the current interest and rapid development of the use of labeled antibodies, it should not be long before the problems of in situ labeling of cell components will be solved. The recent progress in developing these techniques has already brought about many changes that have allowed for increased productivity and clarity. It is hoped that the methodologies described herein can be a further aid in understanding a variety of biological phenomena.

Acknowledgements. The authors wish to thank C. E. TEH for photographic assistance and Ms. DORIS RINGER for typing and editorial work on this manuscript. Portions of the research presented were supported by USPHS Grants GM-00136 and HD-03752, by ERDA Contract AT (45-1)-2225, and by NSF Grant BMS-74-23848.

References

Abermann, R., Salpeter, M. M., Bachmann, L.: High resolution shadowing. In: Principles and Techniques of Electron Microscopy. Hayat, M. A. (ed.). New York: Van Nostrand Reinhold Co., 1972, Vol II, pp. 197 – 217

Averameas, S.: Coupling of enzymes to proteins with glutaraldehyde. Use of the conjugates for detection of antigens and antibodies. Immunochemistry **6**, 43 – 52 (1969)

Averameas, S., Ternynck, T.: Peroxidase labelled antibody and Fab conjugates with enhanced intracellular penetration. Immunochemistry **8**, 1175 – 1179 (1971)

Bolton, A. E., Hunter, W. M.: The labelling of protein to high specific radioactivities by conjugation to a ^{125}I-containing acylating agent. Biochem. J. **133**, 529 – 539 (1973)

Bradley, D. E.: Replica and shadowing techniques. In: Techniques for Electron Microscopy, 2nd ed. Kay, D. (ed.). Springfield, Ill.: Charles C Thomas, 1965, pp. 96 – 152

Branton, D.: Fracture faces of frozen membranes. Proc. Natl. Acad. Sci. U.S. **55**, 1048 – 1056 (1966)

Brown, S., Revel, J. P.: Cell surface labeling for the scanning electron microscope. In: Advanced Techniques in Biological Electron Microscopy II. Koehler, J. K. (ed.). Berlin-Heidelberg-New York: Springer Verlag, 1978, pp. 65 – 88

Budd, G. C.: Recent developments in light and electron microscopic radioautography. Intern. Rev. Cytol. **31**, 21 – 56 (1971)

Bullivant, S.: Freeze-etching and freeze-fracturing. In: Advanced Techniques in Biological Electron Microscopy. Koehler, J. K. (ed.). Berlin-Heidelberg-New York: Springer Verlag, 1973, pp. 67 – 112

Caro, L. G., van Tubergen, R. P.: High resolution autoradiography. I. Methods. J. Cell Biol. **15**, 173 – 188 (1962)

de Petris, S., Raff, M. C.: Normal distribution, patching and capping of lymphocyte surface immunoglobulin studied by electron microscopy. Nature (New Biol.) **241**, 257 – 259 (1973)

Graham, R. C., Jr., Karnovsky, M. J.: The early stages of absorption of injected horseradish peroxidase in the proximal tubules of mouse kidney: Ultrastructural cytochemistry of a new technique. J. Histochem. Cytochem. **14**, 291 – 302 (1966)

Greenwood, F. C., Hunter, W. M., Glover, J. S.: The preparation of ^{131}I labeled human growth hormone of high specific radioactivity. Biochem. J. **89**, 114 – 123 (1963)

Hammerling, U., Polliack, A., Lampen, N., Sabetu, M., de Harven, E.: Scanning electron microscopy of tobacco mosaic virus-labeled lymphocyte surface antigen. J. Exp. Med. **141**, 518 – 523 (1975)

Henderson, W. J., Griffiths, K.: Shadow casting and replication. In: Principles and Techniques of Electron Microscopy. Hayat, M. A. (ed.). New York: Van Nostrand Reinhold Co., 1972, Vol. 2, pp. 151 – 193

Karnovsky, M. J.: A formaldehyde-glutaraldehyde fixative of high osmolarity for use in electron microscopy. J. Cell Biol. **27**, 137 A (1965)

Karnovsky, M. J., Leventhal, M., Unanue, E. R.: Correlated fluorescent and scanning microscopy of immunoglobulin-induced capping and movement of B lymphocytes. J. Cell Biol. **67**, (2 pt. 2), 201 A (1975)

Karnovsky, M. J., Unanue, E. R.: Mapping and migration of lymphocyte surface molecules. Federation Proc. **32**, 55 – 59 (1973)

Karnovsky, M. J., Unanue, E. R., Leventhal, M.: Ligand induced movement of lymphocytes membrane macromolecules. II. Mapping of surface moieties. J. Exp. Med. **136**, 907 – 930 (1972)

Kay, M. M.: Multiple labelling techniques used for kinetic studies of activated human B lymphocytes. Nature (London) **254**, 424 – 426 (1975)

Kinsey, W. H., Koehler, J. K.: Fine structural localization of concanavalin A binding sites on hamster spermatozoa. J. Supramol. Struct. **5**, 185 – 198 (1976)

Kishida, Y., Olsen, B. R., Berg, R. A., Prockop, D. J.: Two improved methods for preparing ferritin-protein conjugates for electron microscopy. J. Cell Biol. **64**, 311 – 339 (1975)

Koehler, J. K.: The freeze-etching technique. In: Principles and Techniques of Electron Microscopy. Hayat (ed.). New York: Van Nostrand Reinhold Co., 1972, Vol. II, pp. 53 – 98

Koehler, J. K.: Changes in antigenic site distribution on rabbit spermatozoa after incubation in "capacitating" media. Biol. Reproduc. **15**, 444 – 456 (1976)

Koehler, J. K., Kinsey, W. H.: Changes in sperm membrane structure during capacitation: A brief review. IIT Research Institute SEM Symposium. Vol. II, p. 325 – 332; O. Johari and R. P. Becker (eds.). Chicago Press Corp.

Koehler, J. K., Perkins, W. D.: Fine structure observation on the distribution of antigenic sites of guinea pig spermatozoa. J. Cell Biol. **60**, 789 – 795 (1974)

Koo, G. C., Stackpole, C. W., Boyse, E. A., Hammerling, U., Lardis, M. D.: Topographical location of H–Y antigen on mouse spermatozoa by immunoelectron microscopy. Proc. Natl. Acad. Sci. U.S. **70**, 1502 – 1505 (1973)

Kraehenbuhl, J. P., Jamieson, J. D.: Solid-phase conjugation of ferritin to Fab-fragments of immunoglobulin G for use in antigen localization on thin sections. Proc. Natl. Acad. Sci. U.S. **69**, 1771 – 1775 (1972)

Mahl, H.: Supermicroscopic determination of the orientation of single aluminum crystals. Metalwirtshaft **19**, 1082 – 1085 (1940)

Marchalonis, J. J.: An enzymic method for the trace iodination of immunoglobulins and other proteins. Biochem. J. **113**, 299 – 305 (1969)

Matter, A., Lisowska-Bernstein, B., Ryser, J. E., Lamelin, J.-P., Vassalli, P.: Mouse thymus-independent and thymus-derived lymphoid cells. II. Ultrastructural studies. J. Expl. Med. **136**, 1008 – 1030 (1972)

Nemanic, M. K., Carter, D. P., Pitelka, D. R., Wofsky, L.: Hapten-sandwich labeling. II. Immunospecific attachment of cell surface markers suitable for scanning electron microscopy. J. Cell Biol. **64**, 311 – 321 (1975)

Nicolson, G. L.: Ultrastructural Localization of Lectin Receptors. In: Advanced Techniques in Biological Electron Microscopy II. Koehler, J. K. (ed.). Berlin-Heidelberg-New York, 1978, pp. 1 – 38

Olsen, B. R., Berg, R. A., Kishida, Y., Prockop, D. J.: Collagen synthesis: Localization of prolyl hydroxylase in tendon cells detected with ferritin-labeled antibodies. Science **182**, 825 – 827 (1973)

Olsen, B. R., Berg, R. A., Kishida, Y., Prockop, D. J.: Further characterization of embryonic tendon fibroblasts and the use of immunoferritin techniques to study collagen biosynthesis. J. Cell Biol. **64**, 340 – 355 (1975)

Olsen, B. R., Prockop, D. J.: Ferritin-conjugated antibodies used for labeling of organelles involved in the cellular synthesis and transport of procollagen. Proc. Natl. Acad. Sci. U.S. **71**, 2033 – 2037 (1974)

Otto, H., Takamiyo, H., Vogt, A.: A two-stage method for cross-linking antibody globulin to ferritin by glutaraldehyde. Antigen binding and agglutinating capacity. J. Immunol. Meth. **3**, 137 – 146 (1973)

Painter, R. G., Tokuyasu, K. T., Singer, S. J.: Immunoferritin localization of intracellular antigens: the use of ultracryotomy to obtain ultrathin sections suitable for direct immunoferritin staining. Proc. Natl. Acad. Sci. U.S. **70**, 1649 – 1653 (1973)

Perkins, W. D., Karnovsky, M. J., Unanue, E. R.: An ultrastructural study of lymphocytes with surface-bound immunoglobulin. J. Exp. Med. **135**, 267 – 276 (1972)

Pinto de Silva, P., Douglas, S. P., Branton, D.: Localization of A-antigen sites on human erythrocyte ghosts. Nature (London) **232**, 194 – 196 (1971)

Rabellino, E., Gray, H. M.: Immunoglobulins on the surface of lymphocytes. III. Bursal origin of surface immunoglobulins on chicken lymphocytes. J. Immunol. **106,** 1418 – 1420 (1971)

Raff, M. C.: Theta isoantigen as a marker of thymus-derived lymphocytes in mice. Nature (London) **224,** 378 – 379 (1969)

Raff, M. C.: Cell surface immunology Sci. Amer. **234,** 30 – 39 (1976)

Raff, M. C., Sternberg, M., Taylor, R. B.: Ig determinants on the surface of mouse lymphoid cells. Nature (London) **225,** 553 – 555 (1970)

Reyes, F., Lejonc, T. L., Gourdin, M. F., Mannoni, P., Dreyfus, B.: The surface morphology of human B lymphocytes as revealed by immunoelectron microscopy. J. Exp. Med. **141,** 392 – 410 (1975)

Reynolds, E. S.: The use of lead citrate at high pH as an electron opaque stain in electron microscopy. J. Cell Biol. **17,** 208 – 213 (1963)

Salpeter, M., Mc Henry, F. A.: Electron microscope autoradiography. Analyses of autoradiograms. In: Advanced Techniques in Biological Electron Microscopy. Koehler, J. K. (ed.). Berlin-Heidelberg-New York: Springer Verlag, 1973, pp. 113 – 152

Singer, S. J.: Preparation of an electron dense antibody conjugate. Nature (London) **183,** 1523 – 1524 (1959)

Smith, S. B., Revel, J. P.: Mapping of concanavalin A binding sites on the surface of several cell types. Develop. Biol. **27,** 434 – 441 (1972)

Takamiya, H., Shimizu, F., Vogt, A.: A two-stage method for cross-linking antibody globulin to ferritin by glutaraldehyde. II. Antigen binding and agglutination capacity. J. Immunol. Meth. **4,** 381 – 386 (1974)

Ukena, T. E., Borysenko, J. Z., Karnovsky, M. J., Berlin, R. D.: Effects of colchicine, cytochalasin B and 2-deoxyglucose on the topographical organization of surface-bound concanavalin A in normal and transformed fibroblasts. J. Cell Biol. **61,** 70 – 82 (1974)

Unanue, E. R., Grey, H. M., Rabellino, E., Campbell, P., Schmidtke, J.: Immunoglobulins on the surface of lymphocytes. II. The bone marrow as the main source of lymphocytes with detectable surface-bound immunoglobulins. J. Exp. Med. **133,** 1188 – 1198 (1971)

Unanue, E. R., Perkins, W. D., Karnovsky, M. J.: Endocytosis by lymphocytes of complexes of anti-Ig with membrane-bound Ig. J. Immunol. **108,** 569 – 572 (1972 a)

Unanue, E. R., Perkins, W. D., Karnovsky, M. J.: Ligand-induced movement of lymphocyte membrane macromolecules. I. Analysis by immunofluorescence and ultrastructural radioautography. J. Exp. Med. **136,** 885 – 906 (1972 b)

van Bruggen, E. F. J., Schniten,V., Wieberiga, E. H., Gruber, M.: Structure and properties of hemocyanins. III. Electron micrographs of hemocyanins from different Gastropoda and Crustacea. J. Mol. Biol. **7,** 249 – 253 (1973)

Wagner, M.: Methods of labeling antibodies for electron microscopic localization of antigens. In: Research in Immunochemistry and Immunobiology. Kwapinski, J. B. G., Day, E. D. (eds.). London: University Park Press. 1973, Vol. 3, pp. 185 – 252

Williams, R. C., Wyckoff, R. W. G.: Applications of metallic shadow casting to microscopy. J. Appl. Phys. **17,** 23 – 33 (1946)

Cell Surface Labeling for the Scanning Electron Microscope

S. S. BROWN and J.-P. REVEL

A. Introduction

The scanning electron microscope (SEM) is a powerful instrument with which to study cell surfaces (WELLS, 1974; REVEL, 1974 a). Compared to other microscopic techniques, it permits the examination of the largest areas of surface at a resolution of better than 10 nm. It is thus particularly well suited to the study of the detailed spatial distribution of specific groups at the cell surface that can be visualized microscopically with appropriate markers.

Although they have only recently been applied to specimens of the SEM (NEMANIC, 1975; REVEL, 1974 b; WELLER, 1974; NEMANIC et al., 1975; BRETTON et al., 1973; BROWN, 1974; HAMMERLING et al., 1975; LINTHICUM et al., 1974; LOBUGLIO et al., 1972; MOLDAY et al., 1975), morphologically recognizable markers have been used to investigate the cell surface for a number of years. Most of the early observations were made with fluorescence labeling at the light microscopic level (COONS, 1956), or with ferritin- (MORGAN, 1972) or enzyme- (AVRAMEAS, 1970) coupled label in thin sections at the electron microscopic level (PEARSE, 1972). The former approach is limited by the resolution of the light microscope, and the latter by the limited amount of surface that can be surveyed per thin section. The amount of surface seen can be increased by preparing sections tangential to the surface, but this is not completely satisfactory, as cell surfaces are never flat enough to remain for long in the plane of section. One way to examine distribution on the entire surface is by serial thin sectioning and reconstruction (STACKPOLE et al., 1971), a very laborious and time-consuming procedure.

To circumvent the problems encountered in sectioning, various transmission electron microscope (TEM) techniques have been devised that maximize the amount of cell surface seen. Thus, NICOLSON and SINGER (1971) lysed cells at an air-water interface, labeled the membranes, then picked them up on an electron microscope grid and observed them directly. The disadvantage of this method is that whole cells in situ cannot be studied. Also, it is not known whether this treatment of the cells would alter labeling in any way. Freeze-etching has also been used to study labeled surfaces (STACKPOLE,

1971; KARNOVSKY et al., 1972; UNANUE et al., 1973). In this technique, ordinarily, the sample is frozen in an antifreeze solution to retard ice crystal formation, broken open, and a replica of the exposed surface is made. This only exposes the interior of membranes, however, and it is necessary to omit the antifreeze to reveal the labeled outer surface by deep etching (PINTO DA SILVA and BRANTON, 1970; TILLACK and MARCHESI, 1970). Large ice crystals can then form, limiting the amount of surface that can be seen and obscuring relations between various surfaces. Instead of freeze-etching, one can prepare replicas of dried samples. This is a good technique for cells attached to and spread on a substratum, as is the case for cell cultures. Cells are fixed, dehydrated, and dried, preferably by the critical point method. Replicas are prepared by coating the sample with platinum and carbon, then digesting away the sample, and viewing the replica in the TEM.

Similar, gold-coated preparations can be viewed in the SEM without digestion. For SEM the beam does not have to pass through the sample, since the image is formed by the secondary electrons escaping from its surface (WELLS, 1974). Use of the SEM thus can avoid replica breakage resulting from handling and drying of the replica during and after digestion, and the loss of those portions of the replica that are hidden from view by the bars of a TEM grid. SEM greatly broadens the range of specimens that can be viewed, since the larger and more complex the surface, the more prone the replica to collapse upon digestion.

Thus it can be seen that scanning microscopy offers distinct advantages for the study of labeled cell surfaces. Its greatest limitation is the relatively poor resolution presently attainable. This resolution however is rapidly improving and may soon rival the 2-nm resolution imposed on TEM replicas by the metals used for shadowing (BROERS et al., 1975; KOMODA and SAITO, 1972).

B. Labeling Techniques for the SEM

Generally speaking, cell surface labeling involves (1) a *label* that binds to a specific molecule or molecules at the cell surface, (2) a *marker* that can be visibly appreciated in the microscope, and (3) a way of attaching the label to the marker. This review will take up each of these subjects in turn, and then turn to a general discussion of problems encountered in cell surface labeling. Labeling techniques have been discussed elsewhere in this volume (NICOLSON). We will focus our discussion on those techniques and problems that are applicable to the SEM.

I. The Label

The criteria for selecting a label are the same for any of the microscopic techniques; the SEM does not place any special constraint on its choice. The nature of the site at the cell surface to be labeled will dictate what may be used as a label. It is desirable to use a label that has as high an affinity and specificity for its binding site as possible. Optimum labeling conditions can best be met, and interpretation is facilitated, if the chemistry of the label and its interaction with the binding site is well understood. A specific inhibitor of the interaction is useful for demonstrating the specificity of the labeling.

If labeling is to be performed on live cells, it must bind under physiologic conditions and not have toxic effects.

Labeling with antibodies is a broadly used approach (PEARSE, 1972; AVRAMEAS, 1970; MORGAN, 1972; COONS, 1956). Here the chemistry is well understood, and specificity can be demonstrated using a variety of immunologic techniques. When appropriate, a hapten inhibitor can be used to remove the label specifically from the cell surface. Extensive purification of the antibody label is probably mandatory. It has been shown to be important to reduce background labeling in thin section (KRAEHENBUHL and JAMIESON, 1972). Whole immunoglobulin can be digested to Fab fragments if a smaller or monovalent label is desired.

Another class of labels is the lectins, which in the last several years have gained in popularity as cell surface labels (NICOLSON, this volume). Originally used in characterizing blood group specificities (BOYD, 1963), lectins of plant or animal origin bind to different sugar moieties, and their interaction at the cell surface can be reversed by a mono- or oligomeric sugar for which they are specific (SHARON and LIS, 1972). Their specificity is not as restricted as that of antibodies.

There are a variety of potential labels that have not been used much for morphologic studies which could be used for cell surface labeling. Affinity labels (BAKER, 1967; SINGER, 1967; SHAW, 1970) that can form covalent linkages and natural labels such as hormones that normally bind to cell surfaces might be useful for microscopy.

II. The Marker

The criteria for selecting markers vary with the microscopic technique used. A marker with a good absorption spectrum or fluorescence emission is used for light microscopy, and one with good electron density and appropriate size, for TEM. For SEM, the shape of the marker also becomes a major criterion, in addition to a size adequate for resolution. The best resolution available with most SEMs toaday is about 10 nm, although a new generation of instruments will probably improve this figure in the near future. By the time this review is published, it is expected that several commercially produced instruments will be capable of resolutions of 3 nm or better. Besides being resolvable the marker should be clearly recognizable in a way suitable for detection with one of the imaging modes of the SEM (secondary, low loss, cathodoluminescence, backscatter, X-ray). Ferritin (diameter 11 nm) illustrates the point that simply being above the limit of resolution of the SEM is not always sufficient for recognition. Widely used as an EM marker (MORGAN, 1972), ferritin is recognized in the TEM by its electron-dense 5.5-nm iron hydroxide core. Imaged in the SEM in the secondary mode, the protein's globular shape does not stand out from other irregularities at the cell surface (REVEL, unpublished). In the following paragraphs we will discuss two types of markers that may be recognizable in the SEM: those that are electron dense and those that have a distinctive shape.

1. *Electron-Dense Markers*

Electron density has been the chief distinguishing characteristic of markers developed for the TEM. In theory, these markers should be distinguishable in the SEM as well. Heavy metals that give rise to scattering and thus contrast in the TEM are sources of the backscattered and secondary electrons that are used for imaging in the SEM. The difficulty with using secondary electrons to visualize an electron-dense marker is that the SEM sample must be coated with heavy metal so that its topography can be studied, and there is thus a problem of contrast between sample and marker. Ways of circumventing this problem, such as using the backscatter mode and X-ray analysis for different metals, have not yet been fully explored (ABRAHAM and DENEE, 1974; MORETON et al., 1974; SUTFIN et al., 1971). One can also conceive of recording the distribution of marker by photographing a sample before coating it with metal and reexamining the same area after coating to correlate marker with surface configuration. Up to now, however, most SEM applications of heavy metal markers have actually used shape as the major criterion for recognizing the marker (WARD et al., 1972; BRETTON et al., 1973).

The electron-dense markers used for SEM were all originally introduced for transmission work. In addition to ferritin, which may come into its own as the resolving power of the SEM improves, attempts have also been made to use enzymes that catalyze the formation of electron-dense products (AVRAMEAS, 1970). NAKANE and PIERCE (1967) coupled peroxidase and acid phosphatase to cell surface labels. Histochemical procedures for these enzymes result in the formation of a lead precipitate for acid phosphatase (ERICSSON and TRUMP, 1965) and of an osmiophilic product for peroxidase (GRAHAM and KARNOVSKY, 1966). There are important theoretical advantages of the enzymes over ferritin for SEM work. For one, the electron density achievable with an enzyme is not solely a function of the number of molecules bound, as would be the case with ferritin, but depends rather on the conditions of the histochemical reaction. The conditions can be varied according to whether increased resolution or sensitivity is desired. One may therefore be able to control more readily the problem of contrast between the sample and the reaction product. Another advantage of the enzymes is their relatively small size. Peroxidase (40,000 daltons) adds relatively little mass to an immunoglobulin label (160,000 daltons), compared to ferritin (750,000 daltons). Thus a peroxidase-coupled antibody can penetrate into a tissue more easily than a ferritin-coupled antibody. This should help in taking advantage of the large specimen size that it is possible to use in the SEM. A peroxidase-coupled reagent is also less likely to be sterically hindered in its interactions with the cell surface than one marked with ferritin. A possible problem with enzyme-based markers is short range migration of the reaction product, which could lower resolution and obscure microheterogeneity in distribution of surface sites. BRETTON et al. (1972a) present evidence against this possibility, using antiferritin coupled to peroxidase.

Peroxidase has been employed in the SEM (WARD et al., 1972; BRETTON et al., 1973). The latter authors see no difference in labeling pattern on L929 cells with concanavalin A (con A) before and after fixation. However, a heterogeneous pattern is found before, and a homogeneous pattern after, fixation when labeling L929 cells with con A and hemocyanin under similar conditions (BROWN, 1974). Thus it may be that differences in arrangement were not distinguished with peroxidase in this case, suggesting that peroxidase may not be an optimum SEM marker.

Another approach to producing an electron-dense marker is autoradiography. This approach in theory has a number of advantages; coupling is unnecessary, and quantitation of the number of sites can be obtained from the specific activity of the radioactive label. A problem is low resolution, which is a function of emulsion grain size and geometry. Using the finest grained emulsion and thin sections, a resolution of under 100 nm is possible in TEM (BACHMANN and SALPETER, 1965). However, the complex geometry of the samples usually used for the SEM can be expected to severely compromise this figure. Emission from the radioactive source in a thin section is most effective in exposing only the overlying silver grains. Since the surface of the SEM sample is not flat, some of the emitted rays can expose silver grains in the emulsion located far from the source of the radioactive decay event.

SEM autoradiography has been attempted (HODGES and MUIR, 1974) but has not been applied to cell surface labeling. The techniques of TEM autoradiography can be used for both TEM and SEM replicas. Samples are coated with emulsion after forming the replica by coating with metal. After exposure and development of the photographic emulsion coating of the replica, the sample must be digested away for the TEM. Care has to be exercised that the silver grains formed during exposure do not dissolve or that the gelatin that holds the silver grains in place is not completely removed. These problems do not arise in SEM since specimen digestion is not required. How-

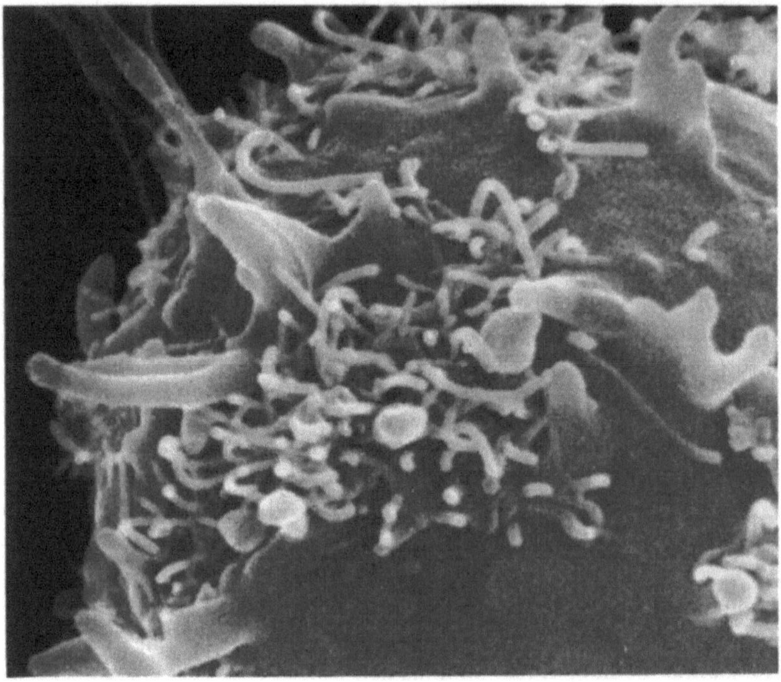

Fig. 1. Tobacco mosaic virus (TMV) as an SEM marker for Thy-1 antigen on a murine lymphocyte, using the hybrid antibody technique. The narrow rod-shaped TMV (sometimes curving) can be distinguished from the broader cell projections. ×29,500. (Hammerling et al., 1975; Fig. 5)

ever, we found that, while readily identifiable in the TEM replica, the silver grains could not be readily seen in the SEM in the secondary emission mode. Hodges and Muir (1974) demonstrated autoradiographic silver grains in the SEM; they point out that their identity can be confirmed by X-ray analysis for silver and that they are readily identifiable in the backscatter mode or in uncoated specimens in the secondary mode, because of their higher backscatter coefficient. Overall resolution is impaired by such techniques, and it becomes difficult to correlate the distribution of silver grains with the topography of the sample.

An important consideration with surface labels for SEM autoradiography is that they should not be taken up inside the cell, since this would give rise to spurious surface labeling. Radioactive membrane precursors would not be suitable for labeling if radioactivity remained in intracellular pools. Autoradiography of thin sections could be done in conjunction with SEM to rule out intracellular sources.

2. *Markers Recognizable by Their Shapes*

The class of marker that is recognizable on the basis of distinctive shape and uniform size has proved more suitable than heavy metal markers for the SEM. Plant viruses were introduced by DE HARVEN's lab as visual markers for TEM. They first employed southern bean mosaic virus (HAMMERLING et al., 1969), a round particle about 25 nm in diameter. KOURILSKY et al. (1971) employed turnip yellow mosaic virus in a study of HL-A antigens. Tobacco mosaic virus, a cylindrical virus with a diameter of 15 nm, that can be 280 – 300 nm long, has also been used (AOKI et al., 1971). This virus can be sonicated into smaller pieces to make a marker of a more useful length and has been used as a marker in the SEM (Fig. 1) (HAMMERLING et al., 1975; NEMANIC et al., 1975). NEMANIC et al. (1975) also used bushy stunt virus (diameter 30 nm) as an SEM marker, and KAY (1975) used T2 (head 70 nm wide) and SV40 (40 – 45 nm).

We have found hemocyanin to be very useful in TEM replicas (SMITH and REVEL, 1972) and SEM studies (REVEL, 1974 a, b; WELLER, 1974; KAY, 1975; NEMANIC et al., 1975). This molecule has a distinctive cylindrical

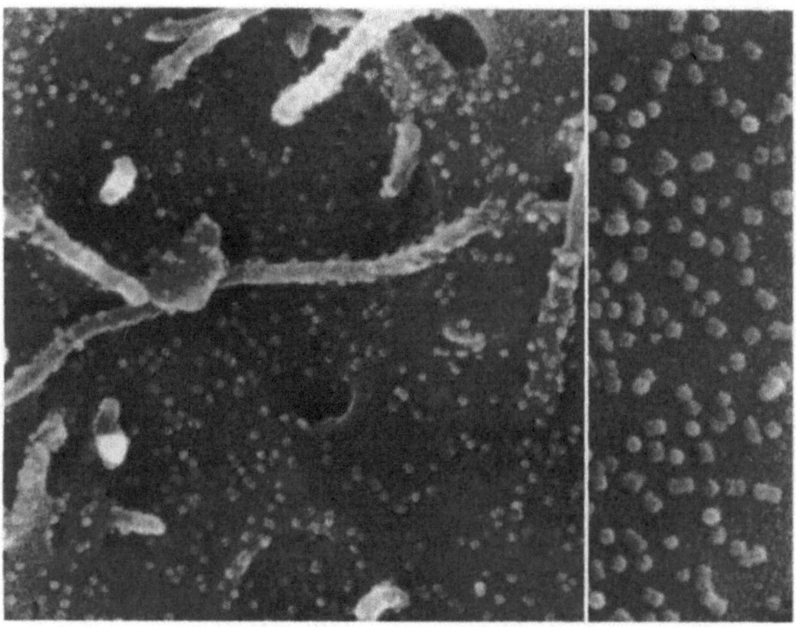

Fig. 2. Hemocyanin as an SEM marker for con A binding sites on a murine tissue culture cell (LA-9), on the left. ×36,000. The picture on the right shows hemocyanin at a higher magnification (×47,500) at which the cylindrical shape of the molecule is more readily appreciated

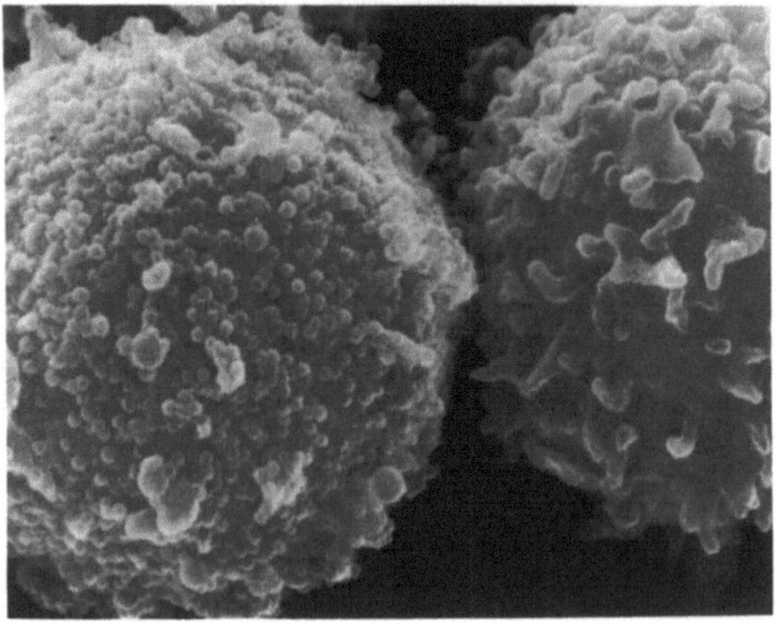

Fig. 3. Latex spheres (10 nm dia.) as an SEM marker for immunoglobulin on murine lymphocytes, using indirect labeling. Only the cell on the left is labeled. ×19,000. (Courtesy of Dr. ROBERT S. MOLDAY, California Institute of Technology)

shape, so that it appears rectangular from the side and circular end-on. The diameter and also mean length is 35 nm, which is comfortably within the limit of resolution of most present-day SEMs (Fig. 2). Like some of the viruses used as markers, it is composed of subunits that dissociate readily under nonphysiologic conditions. These large molecules can be lightly fixed with glutaraldehyde (WOFSY et al., 1974) to prevent this dissociation.

Synthetic markers have been developed for the SEM. MOLDAY et al. (1975) derivatized latex spheres to make them suitable for coupling to antibody labels (Fig. 3). LINTHICUM et al. (1974) and LOBUGLIO et al. (1972) also used latex spheres, relying however on a nonspecific sticking for coupling to label. Spheres are easily recognizable, and any desired size may be selected.

3. *Cathodoluminescent and Other Markers*

Fluorescein would be a very desirable SEM marker, since it has been widely used in light microscopic labeling studies (COONS, 1956) and because of its small size. Fluorescein or other molecules in which fluorescence or cathodoluminescence can be generated could be recognized by using a cathodoluminescence detector available for many makes of SEM (PEASE and HAYES, 1966; DE METS, 1974; FALK et al., 1974; BROECKER et al., 1975; SONI et al., 1975). The resolution of this technique is presently not as good as that of light microscopy. One must also consider the fact that fluorescent molecules are rapidly destroyed in the SEM. A recent report (SPRINGER et al., 1974) suggests that fluorescein can be used in SEM not only by its fluorescence but also because its high ionization potential causes it to be a very good secondary electron emitter. This observation, if it can be extended, may prove to be an exciting development.

III. Coupling Label to Marker

1. Direct Coupling

After label and marker have been selected, the next problem is how to couple the marker to the label. In a few fortunate cases, coupling is not necessary. For instance, con A binds peroxidase (BERNHARD and AVRAMEAS, 1971) and hemocyanin (SMITH and REVEL, 1972) via their sugar residues. Thus con A can bind to the cell surface, and its remaining active sites can then bind the marker molecule. In most cases, however, some sort of coupling procedure is required. Ideally a coupling reagent covalently links one label molecule to one marker molecule, without damaging any of the relevant properties of either.

A widely used coupler is glutaraldehyde, first employed by AVRAMEAS (1969) to couple antibody to peroxidase. Glutaraldehyde can, in principle, form a crosslink between amino groups in different molecules, coupling the molecules together (RICHARDS and KNOWLES, 1968). A drawback is the lack of specificity of this method. Label-label and marker-marker as well as label-marker combinations are possible. Furthermore, some antibodies are inactivated by glutaraldehyde (KRAEHENBUHL et al., 1971). The mechanism of coupling by glutaraldehyde is complex and the precise molecular species involved are not known with certainty, giving rise to much lore about the most appropriate conditions.

Carbodiimide reacts with a carboxyl group, giving a product that can then couple by reacting with an amino group. It seems reasonably innocuous (KENDALL et al., 1971), as long as a carboxyl is not involved in the active site of the label molecule (it inactivates con A, for instance).

SINGER used first m-xylylene diisocyanate (SINGER, 1959), then toluene diisocyanate (SINGER and SCHICK, 1961). The latter was an improvement because its two groups are active under different sets of conditions and it is milder. Difluorodinitrophenylsulfone and bis-diazotized benzidine have also been used (PEARSE, 1972).

2. *Indirect Coupling*

To get around the damage often done to the label and the losses of label and reduction of label affinity incurred during coupling, immunologists employ a variety of indirect methods. Instead of coupling the antibody directly to a marker, one treats cells with an unmodified antibody label (primary antibody) that has been produced, usually, in a rabbit. The cells are then incubated with antirabbit immunoglobulin that has been made in another species and physically coupled to the marker. This latter antibody is generally much easier to obtain than the primary specific antibody. The primary antibody may require a highly purified antigen as inoculum, and much purification and resulting losses by immunoabsorption, whereas the secondary antibody need only be antirabbit and can be commercially obtained. Furthermore, the same antirabbit marker preparation can be used with a variety of rabbit-produced primary antibodies. This avoids having to use separate coupling procedures for each primary antibody used. In addition, indirect techniques provide greater sensitivity than direct techniques because each successive antibody is multivalent.

The hapten sandwich technique (LAMM et al., 1972; WOFSY et al., 1974) is a variation that facilitates double labeling and permits labeling of alloantigens with an indirect technique. A haptenic group is coupled to both label and marker, which can then be associated by an antihapten antibody.

There are variations of the indirect approach that avoid chemical coupling altogether. The 'hybrid antibody' technique (HAMMERLING et al., 1968) is elegant but difficult. It involves making antibody against not only the primary antibody, but also against the marker. These two different antibodies are then digested and reduced to antibody fragments (Fabs) and allowed to recombine so that the resulting reagent has one binding site for the primary antibody and one for the marker. Another approach is the PAP technique of STERNBERGER (1969) and MASON et al. (1969). The primary antibody against a surface antigen is made in a rabbit. Goat antirabbit immunoglobulin then links the primary antibody, which has reacted with the cell surface, to an antibody also made in rabbit, against peroxidase. The peroxidase serves as marker.

3. *Purification and Analysis of Conjugates*

Coupling procedures can produce label-label and marker-marker combinations, in addition to label-marker combinations in various ratios. The reaction mixture can also contain unreacted label or marker. There are a variety of ways to separate out the desired species and to analyze the composition of the species that have been produced. Label, marker, and conjugates can often be separated by column chromatography (AVRAMEAS and TERNYNCK, 1971; KISHIDA et al., 1975), especially if they differ greatly in size. Affinity chromatography is a boon to separation (NICOLSON, this volume), as the species with the activity desired will react with the column, while other species, such as marker-marker combinations or inactivated label, can be washed away. The sugar specificities of a number of lectins correspond to the composition of some commercially available gels, so that con A conjugates can be purified on Sephadex (AGRAWAL and GOLDSTEIN, 1967) and ricin on Agarose (NICOLSON and BLAUSTEIN, 1972). Ammonium sulfate precipitation has been used for purification (AVRAMEAS and TERNYNCK, 1971), as has ultracentrifugation (SINGER and SCHICK, 1961). Separation methods based on density differences can also be used if, for instance, the marker is nonprotein (MOLDAY et al., 1975).

Electrophoresis gives increased resolution in separating species (PIERCE et al., 1964; OTTO et al., 1973), although this technique is more often used analytically (AVRAMEAS and TERNYNCK, 1971; SINGER and SCHICK, 1961; AVRAMEAS, 1969). Immunoelectrophoresis is an excellent technique for analysis (KRAEHENBUHL and JAMIESON, 1972; KISHIDA et al., 1975; AVRAMEAS, 1969), as it can identify which species contain label or marker. This technique can also indicate whether an antibody label is still active after the coupling procedure, as can other standard immunological techniques (BRETTON et al., 1972 a; KRAEHENBUHL and JAMIESON, 1972; KISHIDA et al., 1975). Analytical ultracentrifugation has also been used to analyze conjugates (AVRAMEAS, 1969).

Spectrophotometric methods can be used to obtain the relative concentrations of label and marker in a purified conjugate, if one of these has an absorption maximum where the other does not. For instance, ferritin absorbs well at 410 nm (KISHIDA et al., 1975) and peroxidase at 403 nm (AVRAMEAS and TERNYNCK, 1971). Radioactive labeling of one of the species (NICOLSON, this volume; MOLDAY et al., 1975; KISHIDA et al., 1975) can also help in analyzing the ratio of label to marker in a conjugate.

C. Interpretation of Cell Surface Labeling in the SEM

More than any of the other microscopic techniques available, the SEM provides a way of actually quantitating the number of marker molecules over a large surface area. Although enumeration of marker molecules is relatively easy, inferences about numbers of binding sites should be approached with caution, because of the large size of SEM markers as well as the methods employed to associate them with specific sites. These factors also affect the resolution of the labeling, and whether adequate advantage of the resolution of the SEM is taken. Properties of the sample to be labeled also affect quantitation and resolution, as well as introducing other factors to be taken into consideration. Finally, preparation of the sample for the SEM must not obscure the labeling. These various factors are discussed in turn in the following paragraphs.

I. Quantitation

1. *Influence of Valence of the Label*

A primary consideration in the quantitation of binding sites by SEM is the ratio of label to binding site and marker to label produced by the labeling conditions. The valence of each of these species is important. If there is more than one binding site per surface molecule, that molecule could potentially bind more than one label molecule. Conversely, if the label is multivalent, one label molecule could bind to more than one cell surface molecule, depending on its concentration in solution. This problem would be multiplied in indirect labeling, where there are successive additions of multivalent reagents. Unfortunately, many of the commonly used label molecules are multivalent; antibody molecules are divalent, and different lectins have different valences, all greater than one. Antibodies can be reduced to monovalent fragments (Fab) by reduction of the disulfide bonds that hold them together. There are reports of reducing concanavalin A to functional monovalence (BURGER and NOONAN, 1970; GUNTHER et al., 1973), although the processes involved are not completely understood.

2. *Stoichiometry of the Binding of Label to Marker*

After a monovalent label is obtained, there is a further problem, the ratio of label to marker produced during the coupling reaction. One solution would be to use concentrations of label and marker most favoring a one-to-one combination, then use chromatography or other methods to separate out the desired species. Another approach (KRAEHENBUHL and JAMIESON, 1972) is to bind Fab to an immunoabsorbent column, which protects the Fab's active site as well as immobilizing it so label molecules will not be crosslinked during the coupling. Isocyanate-activated ferritin is then added to the column and the one-to-one conjugate eluted. AVRAMEAS and TERNYNCK (1971) found that they can control the size of their conjugate by adding antibody to peroxidase pretreated with glutaraldehyde (peroxidase is not very reactive with glutaraldehyde, so the percentage of dimers formed during this pretreatment is low). This approach yields primarily a conjugate of one peroxidase to one antibody molecule and some uncoupled antibody, whereas a combination of peroxidase and antibody in the presence of glutaraldehyde produced higher molecular weight material. OTTO et al. (1973) use a similar 'two-step' glutaraldehyde procedure to couple antibody to ferritin, and find less marker aggregation and label inactivation than with the 'one-step' procedure (AVRAMEAS, 1969). KISHIDA et al. (1975) use chemical manipulations to activate marker for reaction with label under conditions that prevent marker-marker combination, obtaining a mostly one-to-one label-marker product.

3. Influence of the Size of the Marker

Even if a strictly univalent reagent is achieved, the large size of most markers for the SEM may sterically preclude a one-to-one correspondence of marker to surface site. Thus, the smaller the reagent, the better its numbers can correspond to the true state of the cell surface. This problem of steric hindrance was demonstrated by BRETTON et al. (1972 a). They showed that labeling with antibody-peroxidase blocks all subsequent labeling with the larger ferritin coupled to the same antibody, whereas the reverse was not true. In addition to steric hindrance between marker molecules, there is the possibility that the cell surface molecules might hinder penetration of large markers, masking the surface sites to be labeled. HUET and BERNADEC (1974) demonstrated that the ratio of con A to peroxidase varies with the cell type labeled. The accessibility of previously attached con A to the marker molecule may vary with steric factors of the cell surface molecules, or it may reflect other differences between different cell types, such as number of active sites of each con A molecule occupied by surface molecules, or mobility of surface sites. Differences in the con A-peroxidase ratio before and after fixation (COLLARD and TEMMINK, 1974) could be explained by the same possibilities.

These difficulties in achieving a one-to-one, or even just a constant ratio between binding site and marker affect more than absolute quantitation; they must also be considered when only a relative quantitation is desired. Unevenness of distribution of binding sites on a cell, and differences between cells could be obscured if the numbers of binding sites being compared are both above the limit imposed by marker steric hindrance, and if labeling conditions are such that both are labeled to that limit.

II. Resolution

1. Size of the Marker

The smaller the marker, and the closer its linkage to the labeled surface site, the better the resolution of the labeling, the limit of course being the capability of the SEM. Fluorescein, with a molecular weight of 500, is the smallest of the markers we have mentioned, and might give the greatest resolution, depending on how closely packed the molecules must be to be imaged in the secondary mode (SPRINGER et al., 1974). Next in size are the enzymes, but here the actual marker is the histochemical reaction product. A one-to-one relation of enzyme and enzyme product cannot be established unless the individual enzyme molecules are far enough apart and reaction conditions are such that products from different molecules do not overlap. BRETTON et al. (1972 a) obtained beads of reaction product 23.5 nm in diameter, which is considerably larger than the size of the enzyme molecule.

The next class of markers contains larger molecules such as hemocyanin (35 nm) and the viruses, which are in the same size range. Synthetic beads can be almost any size chosen, the smallest having a diameter of about 30 nm. All of these markers are easily recognized. They are distinguishable from each other also, so that they could be used in double labeling experiments. (These markers will all appear larger than their actual size in the SEM, due to the addition of a heavy metal coating.)

2. *Size of the Label-Marker Complex*

Another consideration in determining resolution is the length of the linkage between binding site and marker. With direct labeling, this is essentially the diameter of the label molecule, e.g., about 8 nm for tetravalent con A (EDELMAN et al., 1972). An antibody label is about 4 by 25 nm (HALL et al., 1959). Indirect labeling interposes a greater distance between surface site and marker, decreasing resolution. Obviously the ratio in which label is coupled to marker determines the size of the overall reagent and is another factor to be considered for resolution as well as quantitation. Since only the marker is visible in the SEM, it is easy to forget these other factors, which can have a substantial effect on resolution.

III. The Sample

Up to this point, we have examined the effect of various properties of the label and marker on the detection of sites of interest on cell surfaces. We will now turn our attention to those properties of the sample itself, and to steps in sample preparation, that must be considered in evaluating data obtained by SEM-labeling of cell surfaces.

1. *Label-Induced Rearrangements*

A widely encountered phenomenon in cell surface labeling (TAYLOR et al., 1971; COMOGLIO and GUGLIELMONE, 1972; EDIDIN and WEISS, 1972;

SMITH and REVEL, 1972; KOURILSKY et al., 1972; LOOR et al., 1972) is that label can induce a rearrangement of cell surface sites. This phenomenon has been interpreted as support for the idea that individual cell surface molecules can be freely mobile in the plane of the membrane (SINGER and NICOLSON, 1972). Thus the pattern of labeling seen or the relative number of binding sites in different areas of the cell may not correspond to the true distribution before labeling. More subtly, mobility of surface sites may affect quantitative comparisons of labeling, before and after fixation (COLLARD and TEMMINK, 1974) or among different cell types (HUET and BERNADEC, 1974) for instance, by changing the ratio of marker to binding site.

This rearrangement is interesting in itself, but can be avoided in a number of ways. Lowered temperature and metabolic inhibitors eliminate some aspects of rearrangement but do not prevent a small-scale clustering of the label (TAYLOR et al., 1971; SMITH and REVEL, 1972; ROSENBLITH et al., 1973; LOOR et al., 1972; KOURILSKY et al., 1972; EDIDIN and WEISS, 1972; COMOGLIO and GUGLIELMONE, 1972). Prefixation of the sample, with formaldehyde (ROSENBLITH et al., 1973; PARR and OEI, 1973) or glutaraldehyde (COMOGLIO and GUGLIELMONE, 1972; BROWN, 1974; BRETTON et al., 1972 b) often preserves the original distribution, as does very brief incubation with label. To avoid nonspecific labeling after fixation, the precaution is taken of blocking any remaining aldehydes (NEMANIC, 1975) with ammonium chloride, glycine, or sodium borohydride (the latter is very effective, but cannot be used in con A labeling, as it creates specious binding sites) (BROWN, 1974). With prefixation, one runs the risk of inactivating the surface site one wishes to label; many antigens lose their activity after glutaraldehyde treatment (PEARSE, 1972).

2. *Sources of 'False' Labeling*

Fixation can also cause nonspecific labeling if there is inadequate washing beforehand to remove excess label. Specificity can be ascertained by including an appropriate hapten inhibitor or by omitting the label but not marker from the experimental procedure (the latter control is especially used with 'indirect' labeling). Fixation can also cause a misrepresentation of the state of the cell surface if extraneous materials that bind or block the binding of label are not washed off. In studies of label-induced rearrangement, especially, the assumption has often been that the labeled sites are integral to the membrane. However, the material labeled may be adsorbed serum protein or cell secretions that are either more or less specifically associated with the membrane. Freeze-etching has provided us with a few cases (JAN and REVEL, 1974); PINTO DA SILVA and BRANTON, 1970; TILLACK and MARCHESI, 1970) where labeled sites are integral, in addition to conflicting results for other cells (KARNOVSKY and UNANUE, 1973; KARNOVSKY et al., 1972; BROWN,

1974; BENEDETTI et al., 1973; PINTO DA SILVA and MARTINEZ-PALOMO, 1974; GUERIN et al., 1974; MATTER and BONNET,1974; LOOR, 1973).

Labeling of the background (glass or other substrata) has also been observed (BROWN, 1974; BERLIN and UKENA, 1972). This labeling can be both specific (binding, perhaps, to an adsorbed serum protein layer in the case of tissue culture cells) and nonspecific; it cannot be removed by washing. This labeling of the background to which the sample is attached is often more noticeable in specimens prepared for the SEM because so much of the background is visible. While esthetically unpleasant, such background may be salutory, because it forcefully reminds the researcher to ascertain carefully the specificity of the cell surface labeling he is trying to demonstrate. In some cases, background labeling can be circumvented by attaching cells to their substrate after they have been labeled.

3. Types of Samples

The SEM can be used to look at labeled tissues (JAN and REVEL, 1975; MILLER and REVEL, 1975) as well as single cells. There may be artifacts associated with either. Single cells present a problem if they are in suspension; a way must be found to attach them to a substrate. Live cells will often settle and adhere, although their morphology will change from that in the suspended state (DE HARVEN et al., 1975; WETZEL et al., 1974; BROWN, unpublished observations). Prefixation can also cause clumping (DE HARVEN et al., 1975) if the cells are not dilute, as can incubation in multivalent label (thus labeling in suspension to avoid background labeling can cause as many problems as it solves). Cell clumps hinder quantitative observation and are more prone to 'charging' in the SEM (DE HARVEN et al., 1975).

Because of the problems of attachment, one must check for sampling error in a mixed population of cells (one cell type may adhere better than another). WETZEL et al. (1974) and DE HARVEN et al. (1975) have collected cells on filters. WETZEL et al. (1974) get a good yield of prefixed cells if the sample is freeze-dried, but not if it is 'critical point' dried. DE HARVEN et al. (1975) get an almost quantitative yield using live cells and critical point drying.

Collections of cells on glass can be improved by cytocentrifugation (THORNTHWAITE et al., 1975), and also by a variety of treatments of the glass. The most successful, in our hands, has been the use of a polylysine pretreatment (MAZIA et al., 1975). Protamine is cheaper and can also be used to good advantage (STEINHARDT et al., 1971).

Tissues present a problem of penetration, especially to the large SEM markers, unless the surface to be viewed is directly exposed to the labeling solutions. JAN and REVEL (1974) found it necessary to use very long incubation times when labeling retina. Thus the object is to keep the labeling reagents as small as possible when dealing with tissues. For instance, the label could be conjugated with the very small biotin molecule (HEITZMANN and RICHARDS, 1974) and reacted with the tissue. The tissue could then be fixed and dissected before a large marker coupled to avidin could be added. The reaction of biotin with avidin would complete the labeling. Indirect immunologic techniques might also be used.

4. *Subsequent Sample Preparation for the SEM*

After labeling, it is necessary to prepare the sample so as to allow its examination in the SEM (REVEL, 1975). Fixation (with glutaraldehyde, osmium, or both) must be adequate to preserve the label as well as the morphology of the cell. For instance, hemocyanin is notably sensitive to conditions of pH and ionic strength and can deteriorate readily into its component subunits. After fixation, the sample is dehydrated by passing it through a series of acetones or alcohols. The cells must not be allowed to dry out at any time during this procedure, since the attendant flattening and disruption as the sample passes through a liquid-gas interface can often obscure the label (ANDERSON, 1951). Salts must be adequately washed out, or else they may form a layer that obscures cell surfaces or may even crystallize out. The samples should be dried using a technique that minimizes the deleterious effect of surface tension, such as the critical point (ANDERSON, 1951), or freeze drying. In some cases drying from amyl acetate, which has a low surface tension (REVEL and WOLKEN, 1973), may be a quick and adequate technique, even though distortions of the sample are introduced. Finally, careful attention must be paid to coating the sample. Too thick a metal coat can obscure the marker, whereas too little will give a poor signal incompatible with high resolution. We often deposit metal directly on the sample without an intervening carbon coating in order to get as close a replica of the sample as possible.

D. Summary

In this review we have attempted a critical examination of the difficulties that may be encountered in labeling cells for the SEM. There are, as we have seen, difficulties in producing a good stoichiometry between marker and surface sites, partially because of steric factors that are made worse by the large size of most SEM markers. There is the more general and complex issue of modification of the behavior or distribution of sites on the cell surface by the addition of label or by the very treatments designed to circumvent the label's effects. Yet, with due respect to the shortcomings of cell surface labeling, we believe that the SEM, by virtue of the resolution it affords, the many different types of signals it generates, the large size of the specimens that it permits, and the large amount of surface that it reveals, is one of the most ideal tools for the study of cell surface labeling that has been devised.

Acknowledgements. S.S.B. would like to thank Dr. Raymond Teplitz, City of Hope, Duarte, CA., for support (CA-27671) during the writting of this Chapter. This work was supported by USPHS grant to J.P.R., No. GM-06965.

Note added in proof. Since the completion of this review in 1975 there have been a number of important developments in cell surface labeling techniques for the SEM, in particular the use of gold beads by HORISBERGER and ROSSET [Experientia (1976) **32**, 998]. The whole field has been well reviewed by MOLDAY [in SEM: IITRI, 1977, Vol. II. Proc. of the Tenth Annual SEM Symp., O. JOHARI, ed. IIT Res. Inst., Chicago, Ill., p. 59].

References

Abraham, J. L., DeNee, P. B.: Biomedical applications of backscattered electron imaging – one year's experience with SEM histochemistry. In: Scanning Electron Microscopy. Johari, O., Corvin, I. (eds.). Chicago: ITT Research Institute, 1974, pp. 252 – 255

Agrawal, B. B. L., Goldstein, I. J.: Protein-carbohydrate interaction. VI. Isolation of concanavalin A by specific adsorption on cross-linked dextran gels. Biochim. Biophys. Acta **147**, 262 – 271 (1967)

Anderson, T. F.: Techniques for the preservation of three-dimensional structure in preparing specimens for the electron microscope. Trans. N. Y. Acad. Sci. **13**, 130 – 134 (1951)

Aoki, R., Wood, H. A., Old, L. J., Boyse, E. A., de Harven, E., Lardis, M. P., Stackpole, C. W.: Another visual marker of antibody for electron microscopy. Virology **45**, 858 – 862 (1971)

Avrameas, S.: Coupling of enzymes to proteins with glutaraldehyde. Use of the conjugates for the detection of antigens and antibodies. Immunochemistry **6**, 43 – 52 (1969)

Avrameas, S.: Immunoenzyme techniques: enzymes as markers for the localization of antigens and antibodies. Intern. Rev. Cytol. **27**, 349 – 385 (1970)

Avrameas, S., Ternynck, T.: Peroxidase labelled antibody and Fab conjugates with enhanced intracellular penetration. Immunochemistry **8**, 1175 – 1179 (1971)

Bachmann, L., Salpeter, M. M.: Autoradiography with the electron microscope. A quantitative evaluation. Lab. Invest. **14**, 1041 – 1053 (1965)

Baker, B. R.: Design of Active Site Directed Irreversible Enzyme Inhibitors. New York: Wiley 1967

Benedetti, E. L., Dunia, I., Diawara, A.: The organization of the plasma membrane in mammalian cells. Europ. J. Cancer **9**, 263 – 272 (1973)

Berlin, R. D., Ukena, T. E.: Effect of colchicine and vinblastine on the agglutination of polymorphonuclear leucocytes by concanavalin A. Nature (New Biol.) **238**, 120 – 122 (1972)

Bernhard, W., Avrameas, S.: Ultrastructural visualization of cellular carbohydrate components by means of concanavalin A. Exptl. Cell Res. **64**, 232 – 236 (1971)

Boyd, W. C.: The lectins: their present status. Vox Sang. **8**, 1 – 32 (1963)

Bretton, R., Clark, D. A., Nathanson, L.: The cytochemical detection of concanavalin-A binding on cell surface by scanning electron microscopy. J. Microscopie **17**, 93 – 96 (1973)

Bretton, R., Ternynck, T., Avrameas, A.: Comparison of peroxidase and ferritin labelling of cell surface antigens. Exptl. Cell Res. **71**, 145 – 155 (1972 a)

Bretton, R., Wicker, R., Bernhard, W.: Ultrastructural localization of concanavalin A receptors in normal and SV 40 transformed hamster and rat cells. Int. J. Cancer **10**, 397 – 410 (1972 b)

Broecker, W., Schmidt, E. H., Pfefferkorn, G., Beller, F. K.: Demonstration of cathodoluminescence in fluorescein marked biological tissues. In: Scanning Electron Microscopy. Johari, O., Corvin, I. (eds.). Chicago: ITT Research Institute, 1975, pp. 243 – 249

Broers, A. N., Panessa, B. J., Gennaro, J. F.: High resolution SEM of biological specimens. In: Scanning Electron Microscopy. Johari, O., Corvin, I. (eds.). Chicago: ITT Research Institute, 1975, pp. 234 – 239

Brown, S. S.: Distribution of label on the cell surface. Ph. D. Thesis, Harvard University, 1974

Burger, M. M., Noonan, K. D.: Restoration of normal growth by covering of agglutinin sites on tumor cell surfaces. Nature (London) **228**, 512 – 515 (1970)

Collard, J. G., Temmink, J. H. M.: Binding and cytochemical detection of cell-bound concanavalin A. Exptl. Cell Res. **86**, 81 – 86 (1974)

Comoglio, P. M., Guglielmone, R.: Two dimensional distribution of concanavalin A receptor molecules on fibroblast and lymphocyte plasma membranes. FEBS Lett. **27**, 256 – 258 (1972)

Coons, A. H.: Histochemistry with labeled antibody. Intern. Rev. Cytol. **5**, 1 – 23 (1956)

De Harven, E., Lampen, N., Polliack, A., Warfel, A., Fogh, J.: New observations on methods for preparing cell suspensions for scanning electron microscopy. In: Scanning Electron Microscopy. Johari, O., Corvin, I. (eds.). Chicago: ITT Research Institute, 1975, pp. 362 – 365

De Mets, M.: Cathodoluminescence of organic chemicals. In: Principles and Techniques of Scanning Electron Microscopy. Hayat, M. A. (ed.). New York: Van Nostrand Reinhold, 1974, Vol. 2, pp. 1 – 20

Edelman, G. M., Cunningham, B. A., Reeke, G. N., Becker, J. W., Waxdal, M. J., Wang, J. L.: The covalent and three-dimensional structure of concanavalin A. Proc. Natl. Acad. Sci. U. S. **69**, 2580 – 2584 (1972)

Edidin, M., Weiss, A.: Antigen cap formation in cultured fibroblasts: a reflection of membrane fluidity and of cell motility. Proc. Natl. Acad. Sci. U. S. **69**, 2456 – 2459 (1972)

Ericsson, J. L. E., Trump, B. F.: Observations on the application to electron microscopy of the lead phosphate technique for the demonstration of acid phosphatase. Histochemie **4**, 470 – 487 (1965)

Falk, R. H., Bayer, D. E., Hess, F. D.: Spatial relationships of pesticides on leaf surfaces. In: Scanning Electron Microscopy. Johari, O., Corvin, I. (eds.). Chicago: ITT Research Institute, 1974, pp. 429 – 435

Graham, R. C., Karnovsky, M. J.: The early stages of absorption of injected horseradish peroxidase in the proximal tubules of mouse kidney: ultrastructural cytochemistry by a new technique. J. Histochem. Cytochem. **14**, 291 – 302 (1966)

Guerin, C., Zachowski, A., Prigent, B., Paraf, A., Dunia, I., Diawara, M. A., Benedetti, E. L.: Correlation between the mobility of inner plasma membrane structure and agglutination by concanavalin A in two cell lines of MOPC 173 plasmocytoma cells. Proc. Natl. Acad. Sci. U. S. **71**, 114 – 117 (1974)

Gunther, G. R., Wang, J., Yahara, I., Cunningham, B. A., Edelman, G. M.: Concanavalin A derivatives with altered biological properties. Proc. Natl. Acad. Sci. U. S. **70**, 1012 – 1016 (1973)

Hall, C. E., Nisonoff, A., Slayter, H. S.: Electron microscopic observations of rabbit antibodies. J. Biophys. Biochem. Cytol. **6**, 407 – 412 (1959)

Hammerling, U., Aoki, T., de Harven, E., Boyse, E. A., Old, L. J.: Use of hybrid antibody with anti-γG and anti-ferritin specificities in locating cell surface antigens by electron microscopy. J. Exp. Med. **128**, 1461 – 1473 (1968)

Hammerling, U., Aoki, T., Wood, H. A., Old, L. J., Boyse, E. A., de Harven, E.: New visual markers of antibody for electron microscopy. Nature (London) **223**, 1158 – 1159 (1969)

Hammerling, U., Polliack, A., Lampen, N., Sabety, M., de Harven, E.: Scanning electron microscopy of tobacco mosaic virus-labeled lymphocyte surface antigens. J. Exp. Med. **141**, 518 – 523 (1975)

Heitzmann, H., Richards, F. M.: Use of the avidin-biotin complex for specific staining of biological membranes in the electron microscope. Proc. Natl. Acad. Sci. U. S. **71**, 3537 – 3541 (1974)

Hodges, G. M., Muir, M. D.: Autoradiography of biological tissues in the scanning electron microscope. Nature (London) **247**, 383 – 385 (1974)

Huet, C., Bernadec, A.: Peroxidase binding to cell-bound concanavalin A. Exptl. Cell Res. **89**, 429 – 431 (1974)

Jan, L. Y., Revel, J.-P.: Ultrastructural localization of rhodopsin in the vertebrate retina. J. Cell Biol. **62**, 257 – 273 (1974)

Jan, L. Y., Revel, J.-P.: Hemocyanin-antibody labeling of rhodopsin in mouse retina for a scanning electron microscope study. J. Supramol. Struc. **3**, 61 – 66 (1975)

Karnovsky, M. J., Unanue, E. R.: Mapping and migration of lymphocyte surface macromolecules. Federation Proc. **32**, 55 – 59 (1973)

Karnovsky, M. J., Unanue, E. R., Leventhal, M.: Ligand-induced movement of lymphocyte membrane macromolecules. II Mapping of surface moieties. J. Exp. Med. **136**, 907 – 930 (1972)

Kay, M. M. B.: Multiple labeling technique used for kinetic studies of activated human B lymphocytes. Nature (London) **254**, 424 – 426 (1975)

Kendall, P. A., Polak, J. M., Pearse, A. E. G.: Carbodimide fixation for immunohistochemistry: observations on the fixation of polypeptide hormones. Experientia **27**, 1104 – 1106 (1971)

Kishida, Y., Olsen, B. R., Berg, R. A., Prockop, D. J.: Two improved methods for preparing ferritin-protein conjugates for electron microscopy. J. Cell Biol. **64**, 331 – 339 (1975)

Komoda, T., Saito, S.: Experimental resolution limit in the secondary electron mode for a field emission source scanning electron microscope. In: Scanning Electron Microscopy. Johari, O., Corvin, I. (eds.). Chicago: ITT Research Institute, 1912, pp. 129 – 135

Kourilsky, F. M., Silvestre, D., Neauport-Sautes, C., Dausset, J., Levy, J. P.: Ultrastructural localization of HL-A antigens at cell surface. Transplantation Proc. **3**, 1203 – 1207 (1971)

Kourilsky, F. M., Silvestre, D., Neauport-Sautes, C., Loosfelt, Y., Dausset, J.: Antibody-induced redistribution of HL-A antigens at the cell surface. Europ. J. Immunol. **2**, 249 – 257 (1972)

Kraehenbuhl, J. P., De Grandi, P. B., Campiche, M. A.: Ultrastructural localization of intracellular antigen using enzyme-labeled antibody fragments. J. Cell Biol. **50**, 432 – 445 (1971)

Kraehenbuhl, J. P., Jamieson, J. D.: Solid-phase conjugation of ferritin to Fab-fragments of immunoglobulin G for use in antigen localization on thin sections. Proc. Natl. Acad. Sci. U. S. **69**, 1771 – 1775 (1972)

Lamm, M. E., Koo, G. C., Stackpole, C. W., Hammerling, U.: Hapten-conjugated antibodies and visual markers used to label cell surface antigens for electron microscopy: an approach to double labeling. Proc. Natl. Acad. Sci. U. S. **69**, 3732 – 3736 (1972)

Linthicum, D. S., Sell, S., Wagner, R. M., Trefts, P.: Scanning immunoelectron microscopy of mouse B and T lymphocytes. Nature (London) **252**, 1973 – 1976 (1974)

LoBuglio, A. F., Rinehart, J. J., Balcerzak, S. P.: A new immunologic marker for scanning electron microscopy. In: Scanning Electron Microscopy. Johari, O., Corvin, I. (eds.) Chicago: ITT Research Institute, 1972, pp. 313 – 319

Loor, F.: Lymphocyte membrane particle redistribution induced by a mitogenic/capping dose of the phytohemagglutinin of *Phaseolus vulgaris*. Europ. J. Immunol. **3**, 112 – 116 (1973)

Loor, F., Forni, S., Pernis, B.: The dynamic state of the lymphocyte membrane. Factors affecting the distribution and turnover of surface immunoglobulins. Europ. J. Immunol. **2**, 203 – 212 (1972)

Mason, T. E., Phifer, R. F., Spicer, S. S., Swallow, R. A., Dreskin, R. B.: An immunoglobulin-enzyme bridge method for localizing tissue antigens. J. Histochem. Cytochem. **17**, 563 – 569 (1969)

Matter, A., Bonnet, C.: Effect of capping on the distribution of membrane particles in thymocyte membranes. Europ. J. Immunol. **4**, 704 – 707 (1974)

Mazia, D., Schatten, G., Sale, W.: Adhesion of cells to surfaces coated with polylysine. Applications to electron microscopy. J. Cell Biol. **66**, 198 – 200 (1975)

Miller, M. M., Revel, J.-P.: Scanning electron microscopy of epithelia prepared by blunt dissection. Anat. Rec. **183**, 339 – 358 (1975)

Molday, R. S., Dreyer, W. J., Rembaum, A., Yen, S. P. S.: New immunolatex spheres: visual markers of antigens on lymphocytes for scanning electron microscopy. J. Cell Biol. **64**, 75 – 88 (1975)

Moreton, R. B., Echlin, P., Gupta, B. L., Hall, T. A., Weisfogh, T.: Preparation of frozen hydrated tissue sections for x-ray microanalysis in the scanning electron microscope. Nature (London) **247**, 113 – 115 (1974)

Morgan, C.: The use of ferritin-conjugated antibodies in electron microscopy. Intern. Rev. Cytol. **32**, 291 – 326 (1972)

Nakane, P. K., Pierce, G. B.: Enzyme-labeled antibodies for the light and electron microscopic localization of tissue antigens. J. Cell Biol. **33**, 307 – 318 (1967)

Nemanic, M.: On cell surface labeling for the SEM. In: Scanning Electron Microscopy. Johari, O., Corvin, I. (eds.). Chicago: ITT Research Institute, 1975, pp. 342 – 349

Nemanic, M. K., Carter, D. P., Pitelka, D. R., Wofsy, L.: Hapten-sandwich labeling. II. Immuno-specific attachment of cell surface markers suitable for scanning electron microscopy. J. Cell Biol. **64**, 311 – 312 (1975)

Nicolson, G. L., Blaustein, J.: The interaction of *Ricinis communis* agglutinin with normal and tumor cell surfaces. Biochim. Biophys. Acta **266**, 543 – 547 (1972)

Nicolson, G. L., Singer, S. J.: Ferritin-conjugated plant agglutinins as specific saccharide stains for electron microscopy: application to saccharides bound to cell membranes. Proc. Natl. Acad. Sci. U. S. **68**, 942 – 945 (1971)

Otto, H., Takamiya, H., Vogt, A.: A two-stage method for crosslinking antibody globulin to ferritin by glutaraldehyde. Comparison between the one-stage and the two-stage method. J. Immunol. Meth. **3**, 137 – 146 (1973)

Parr, E. L., Oei, J. S.: Immobilization of membrane H-2 antigens by paraformaldehyde fixation. J. Cell Biol. **59**, 537 – 542 (1973)

Pearse, A. G. E.: Histochemistry, Theoretical and Applied. Baltimore: Williams and Wilkins 1972

Pease, R. F. W., Hayes, T. L.: Scanning electron microscopy of biological material. Nature (London) **210**, 1049 (1966)

Pierce, G. B., Ram, J. S., Midgley, A. R.: The use of labeled antibodies in ultrastructural studies. Int. Rev. Exp. Path. **3**, 1 – 34 (1964)

Pinto da Silva, P., Branton, D.: Membrane splitting in freeze etching. Covalently bound ferritin as a membrane marker. J. Cell Biol. **45**, 598 – 605 (1970)

Pinto da Silva, P., Martinez-Palomo, A.: Induced redistribution of membrane particles, anionic sites and con A receptors in *Entameoba histolytica.* Nature (London) **249**, 170 – 171 (1974)

Revel, J.-P.: Scanning electron microscope studies of cell surface and morphology and labeling, *in situ* and *in vitro.* In: Scanning Electron Microscopy. Johari, O., Corvin, I. (eds.). Chicago: ITT Research Institute, 1974 a, pp. 541 – 547

Revel, J.-P.: Contacts and junctions between cells. Transport at the cellular level. Symp. Soc. Exp. Biol. **27**, 447 – 461 (1974 b)

Revel, J.-P.: Elements of scanning electron microscopy for biologists. In: Scanning Electron Microscopy. Johari, O., Corvin, I. (eds.). Chicago: ITT Research Institute, 1975, pp. 687 – 696

Revel, J.-P., Wolken, K.: Electronmicroscope investigations of the underside of cells in culture. Exp. Cell Res. **78**, 1 – 14 (1973)

Richards, F. M., Knowles, J. R.: Glutaraldehyde as a protein cross-linking reagent. J. Mol. Biol. **37**, 231 – 233 (1968)

Rosenblith, J. Z., Ukena, T. E., Yin, H. H., Berlin, R. D., Karnovsky, M. J.: A comparative evaluation of the distribution of concanavalin A binding sites on the surfaces of normal, virally-transformed, and protease-treated fibroblasts. Proc. Natl. Acad. Sci. U. S. **70**, 1625 – 1629 (1973)

Sharon, N., Lis, H.: Lectins, cell-agglutinating and sugar specific proteins. Science **77**, 949 – 959 (1972)

Shaw, E.: Selective chemical modification of proteins. Physiol. Rev. **50**, 244 – 296 (1970)

Singer, S. J.: Preparation of an electron-dense antibody conjugate. Nature (London) **183**, 1523 – 1524 (1959)

Singer, S. J.: Covalent labeling of active sites. Advan. Protein Chem. **22**, 1 – 54 (1967)

Singer, S. J., Nicolson, G. L.: The fluid mosaic model of the structure of cell membranes "fluid mosaic model." Science **175**, 720 – 731 (1972)

Singer, S. J., Schick, A. F.: The preparation of specific stains for electron microscopy prepared by the conjugation of antibody molecules with ferritin. J. Biophys. Biochem. Cytol. **9**, 519 – 537 (1961)

Smith, S. B., Revel, J.-P.: Mapping of concanavalin A binding sites on the surface of several cell types. Develop. Biol. **27**, 434 – 441 (1972)

Soni, S. L., Kalnins, V. I., Haggis, G. H.: Localization of caps on mouse B lymphocytes by scanning electron microscopy. Nature (London) **255**, 717 – 719 (1975)

Springer, E. L., Riggs, J. L., Hackett, A. J.: Viral identification of scanning electron microscopy of preparations stained with fluorescein-labeled antibody. J. Virol. **14**, 1623 – 1626 (1974)

Stackpole, C. W.: Topography of cell surface antigens. Transplantation Proc. **3**, 1199 – 1201 (1971)

Stackpole, C. W., Aoki, T., Boyse, E. A., Old, L. J., Lumley-Frank, J., de Harven, E.: Cell surface antigens: serial sectioning of single cells as an approach to topographical analysis. Science **172**, 472 – 474 (1971)

Steinhardt, R. A., Lundin, L., Mazia, D.: Bioelectric responses of the echinoderm egg to fertilization. Proc. Natl. Acad. Sci. U. S. **68**, 2426 – 2430 (1971)

Sternburger, L. A.: Some new developments in immunocytochemistry. Mikroskopie **25**, 346 (1969)

Sutfin, L. V., Holtrop, M. E., Ogilvie, R. E.: Microanalysis of individual mitochondrial granules with diameters less than 1000 Angstroms. Science **174**, 947 – 949 (1971)

Taylor, R. B., Duffus, W. P. H., Raff, M. C., dePetris, S.: Redistribution and pinocytosis of lymphocyte surface immunoglobulin molecules induced by anti-immunoglobulin antibody. Nature (New Biol.) **233**, 225 – 229 (1971)

Thornthwaite, J. T., Thornthwaite, B. N., Cayer, M. L., Hart, M. A., Leif, R. C.: A new method for preparing cells for critical point drying. In: Scanning Electron Microscopy. Johari, O., Corvin, I. Chicago: ITT Research Institute, 1975, pp. 387 – 392

Tillack, T. W., Marchesi, V. T.: Demonstration of the outer surface of freeze-etched red blood cell membranes. J. Cell Biol. **45**, 649 – 653 (1970)

Unanue, E. R., Karnovsky, M. J., Engers, H. D.: Ligand-induced movement of lymphocyte membrane macromolecules. III. Relationship between the formation and fate of anti-Ig-surface Ig complexes and cell metabolism. J. Exp. Med. **137**, 675 – 689 (1973)

Ward, H. A., Yamana, S., Phil, E., Nairn, R. C.: Ultrastructural localization of anti-lymphocyte globulin on viable lymphocytes by immunoperoxidase tracing. Immunology **23**, 61 – 68 (1972)

Weller, N. K.: Visualization of concanavalin A-binding sites with scanning electron microscopy. J. Cell Biol. **63**, 699 – 707 (1974)

Wells, O. C.: Scanning Electron Microscopy. New York: McGraw Hill 1974

Wetzel, B., Canon, G. B., Alexander, E. L., Erickson, B. W., Jr., Westbrook, E. W.: A critical approach to the scanning electron microscopy of cells in suspension. In: Scanning Electron Microscopy. Johari, O., Corvin, I. (eds.). Chicago: ITT Research Institute, 1974, pp. 581 – 587

Wofsy, L., Baker, P. C., Thompson, K., Goodman, J., Kimura, J., Henry, C.: Hapten-sandwich labeling. I. A general procedure for simultaneous labeling of multiple cell surface antigens for fluorescence and electron microscopy. J. Exp. Med. **140**, 523 – 537 (1974)

Low-Temperature Biological Scanning Electron Microscopy

P. ECHLIN

A. Introduction

Nearly all biological material is soft and contains a high proportion of water, which means that the mechanical strength of cells and tissues must be substantially increased before their internal contents can be revealed in a controlled and ordered way. The presence of water in biological material also means that, with a few exceptions, it cannot be examined while in the hydrated state in any electron beam instrument, and although there are now a large number of techniques that dehydrate and strengthen cells and tissues, they do so only at the expense of the specimen. Without exception the chemical agents used to achieve this stabilization, dehydration, and strengthening are toxic to living cells. A list of such chemicals would include strong oxidants and reducing agents, non-polar fluids, carcinogens, heavy metals, and polar dehydrating agents.

Low-temperature techniques that rely largely on a physical rather than on a chemical approach can also stabilize and dehydrate cells and tissues but without many of the deleterious side effects associated with the commonly used wet chemical methods. Biological material may be frozen and the water in the cell fluids either retained as ice and the cells examined at low temperature, replicas made of the frozen surfaces for subsequent examination at ambient temperatures, or the frozen cell water removed by sublimation and the material embedded in wax or plastic in order that it may be fractured or sectioned.

The low-temperature conversion of liquid water to ice and the solidification of non-aqueous components increase the mechanical stability of the biological material allowing the internal contents to be revealed. The substantial increase in the viscosity of water that is brought about by freezing also helps to immobilize the soluble constituents of the cells, allowing them to be analysed in situ.

In devising and using any preparative technique in biology we should endeavour to introduce the minimum perturbation into the system being examined. Low-temperature preparation techniques offer one of the most promising — and frustrating — routes to follow, and it is the purpose of this paper to examine how far we are towards achieving this goal with particular regard to examining specimens in the Scanning Electron Microscope (SEM). It is proposed here to consider these techniques in relation to SEM, although much of what will be discussed is common to all forms of electron microscopy and high-energy beam-mediated analytical procedures. Early papers (ECHLIN, 1973; 1975 a) have dealt with some of the instrumental aspects and application of low temperature studies to SEM.

B. Low-Temperature Solidification of Cell and Tissue Fluids

The title for this Section is not a deliberate obfuscation, but a genuine attempt to express the initial act of what is commonly called 'freezing' in more scientific terms. The term freezing is commonly used to describe the process by which liquid water is converted to crystalline ice. In most biological systems when one progressively lowers the temperature of the environment below ambient this is exactly what occurs, and unless great care is taken, usually with disasterous consequences. Water only rarely exists as such in biological tissues, and most systems contain a mixture of aqueous salt solutions and dilute colloids. When the temperature of such mixtures is lowered a phase separation occurs with the formation of a pure ice crystal within a liquid salt-water phase. As the crystallization proceeds, so the liquid phase contains an increasing concentration of ions and electrolytes, which is also eventually converted to a solid phase consisting of ice crystals and hydrated salts. The temperature at which a salt solution can no longer exist in liquid form is known as the eutectic point. Although the eutectic points of many pure salts in pure water are known, it would be impossible to calculate the eutectic point of cell and tissue fluids because of the large number of different dissolved salts involved and the complex interaction of such salts with organic macromolecules in the cell matrix. However, it is generally considered that the phase separation in biological systems proceeds to about 253° K. During phase separation there is a considerable re-location of ions and electrolytes, with the consequent danger either of denaturation of biopolymers or irreversible cell injury due to the osmotic reduction of cell volume (MERYMAN, 1974). A further complication can occur when ice-crystallization occurs outside cells or between cells in a multi-cellular tissue. Such ice crystal growth is usually at the expense of cellular water with the consequence that the cells are progressively dehydrated. As the extra-cellular ice developes, the extra-cellular solutes are concentrated producing osmotic gradients across the cell membrane.

In nearly all cases the phase separation, denaturation potential, and osmotic effects are exacerbated by slow freezing. Under some circumstances, the cell water does not solidify at all or may be transformed either to an amorphous glass or to microcrystalline ice. The cell water that does not solidify usually constitutes between 10 – 15% of the total cell water and is the water that is bound tightly enough to hydrophilic molecules to make it unavailable for ice crystal growth. There is considerable doubt whether amorphous or vitrified water can be formed in biological systems because of the very high cooling rates required to achieve this state. Rapid cooling may be achieved either with 10 – 20 μm microdroplets (BACHMANN and SCHMITT, 1971; PLATTNER et al., 1972) with small pieces of biological material embedded between thin layers of water (TAYLOR and GLAESER, 1973; HEIDE and GRUND, 1974; TAYLOR, 1975; TAYLOR and GLAESER, 1976) or when the critical freezing rate of water is lowered by the application of high pressure (RIEHLE, 1968 a).

A great deal of work has already been carried out during the past 30 years elucidating the process of freezing in cells and tissues, both in relation to preservation of vitality and for ultrastructural detail. Reference should be made to the recent papers by REBHUN (1972), MOOR (1973), MACKENZIE (1972), and MERYMAN (1974), which are reviews of the progress that has been made and are an excellent source for the secondary literature on the subject.

There is general agreement among biologists using low-temperature techniques that our specimen preparation techniques should aim at minimizing ice-crystal formation, and we will now consider the ways by which we may achieve this goal in specimens to be examined in the SEM.

C. Pre-treatment Before the Cooling Process

The pre-treatment falls into five main categories: chemical fixation, artificial nucleation, the use of cryoprotectants, embedding agents, and non-chemical pre-treatment.

I. Chemical Fixation

As recently as 1973, Alan Boyde and I (BOYDE and ECHLIN, 1973) recommended that some form of chemical fixation would always be useful in morphologic studies in the SEM. Although this advice is still valid for most specimens, one cannot be unimpressed by the high fidelity of structure in replicas of freeze-etched specimens and to a lesser extent in the SEM pictures of frozen-fractured specimens, both of which have been prepared without the use of fixatives. The singular advantage to be gained from using fixatives is that they cross-link and to a certain extent stabilize macromolecules. They do not prevent ice crystal damage and can cause soluble constituents to leak from the cells. DORGE et al. (1975) showed that the Na^+ and K^+ concentrations in frog skin were severely altered following fixation in glutaraldehyde, whereas the P concentration was only slightly changed. These and other findings would indicate that, depending on the nature of the soluble constituent, some degree of fixation may be permitted. Unfortunately there have been very few comprehensive systematic studies on the effect of pre-fixation on the morphology of frozen specimens or on the local concentration of soluble constituents in cells. A more recent paper (HOLBROOK et al., 1976) showed that even the most mild fixation can cause a substantial loss of soluble constituents from cells. For the moment it behoves any experimenter who is concerned with analysis and morphology to be prudent in the use of these materials.

II. Artificial Nucleators

One important factor determining the size of ice crystals is the rate of formation of crystal nuclei of a critical size — the so-called nucleation rate. The higher the nucleation rate, the smaller will be the ice crystals that are formed. Because it is not presently practical to vitrify small pieces of biological material, it would seem appropriate to settle for the next best thing and have a specimen in which the water is converted to a large number of very small crystals. Since ice can form around nuclei of foreign material (heterogeneous nucleation) it may be useful to infiltrate the specimen before freezing with artificial nucleating agents in order to accelerate the formation of many small crystals. BOYDE (1974) recommended using a 1% chloroform in water solution, which it is claimed increases the nucleation rate. Unfortunately chloroform is toxic to cells and undoubtedly causes changes in membrane permeability. In any case it is doubtful whether artificial nucleators would be of much use in the cytoplasm of cells, because there are already a large number of hydrophilic molecules and minute inclusions in the cytoplasm that would act as nucleating sites for ice crystal formation. It should be noted, however, that it is unlikely that the bound water would be in a state to bring about homogeneous nucleation. It would, however, seem useful to further explore the use of artificial nucleators to accelerate ice crystal formation in the large, highly aqueous vacuoles and inter-cellular spaces of mature plant cells, and the tissue lumens of some animal cells.

III. Cryoprotection

It has been known for some time that certain chemicals can protect cells and tissues from the disruptive action of freezing and thawing. Cryoprotectants fall into two main classes: (1) those, generally of low molecular weight, that penetrate the cells, and include chemicals such as glycerol, sucrose, dimethylsulfoxide, ethylene glycol, methanol, and even glucose, and (2) high-molecular-weight compounds such as polyhydric alcohols like sorbitol and mannitol, dextran, polyvinylpyrrolidone (PVP), and hydroxymethyl starches that do not penetrate the cells.

The mode of action of cryoprotectants is not clearly understood. They all have a strong affinity for water and may well act by binding water molecules sufficiently strongly to make them unavailable for ice crystal growth. In this connection it could be argued that the cryoprotectants may be causing incipient dehydration of the cells and that the images we see may not be representative of the naturally hydrated cell. The ability of the low-molecular-weight polyhydric alcohols such as glycerol to prevent freezing damage in cells is believed to lie in their colligative properties and in the ability to act as a solvent, thereby keeping potentially harmful salts in solution as they are concentrated during ice formation. The presence of penetrating cryoprotectives inside cells can also lower the temperature at which crystallization starts and diminish the rate of ice crystal growth.

Although excellent morphologic preservation has been obtained from tissues impregnated with penetrating cryoprotectives, their presence may also introduce artefacts. Some, i. e., methanol and glycerol, are toxic to cells, others such as glycerol and dimethylsulfoxide can affect membrane permeability (see ROBARDS, 1974). Recent studies by SJOSTROM and THORNELL (1976) have shown that the majority of soluble elements can be lost even from frozen sections when they are cut on a cryoprotective agent such as dimethylsulfoxide maintained at low temperatures. All of the penetrating cryoprotectants can, if care is not taken with the infiltration procedure, exert osmotic effects on cells. Another major problem with many penetrating cryoprotectants is that they are difficult to remove from cells by physical means, which precludes their use in specimens that are frozen-dried and examined in the SEM.

Studies in progress at Cambridge show that non-penetrating cryoprotectives do not appear to suffer from many of these disadvantages although the evidence is variable and inconclusive. Studies by FRANKS and SKAER (1976) have shown that buffered solutions of PVP effectively prevent ice crystal damage in insect nervous tissue. These studies have now been extended to a wide range of plant and animal tissues using a variety of large-molecular-weight cryoprotectants. The cryoprotectant activity, and the pyhsicochemical, morphologic, and physiologic effects of these substances on cells are reported elsewhere in a series of papers by FRANKS et al., 1977; ECHLIN et al., 1977; and SKAER et al., 1977. More work is urgently needed on the mode of action of non-penetrating cryoprotectives on biological material. The non-penetrating cryoprotectives probably act in much the same way as their penetrating counterparts by binding water molecules and diminishing ice crystal formation and growth. This in turn would

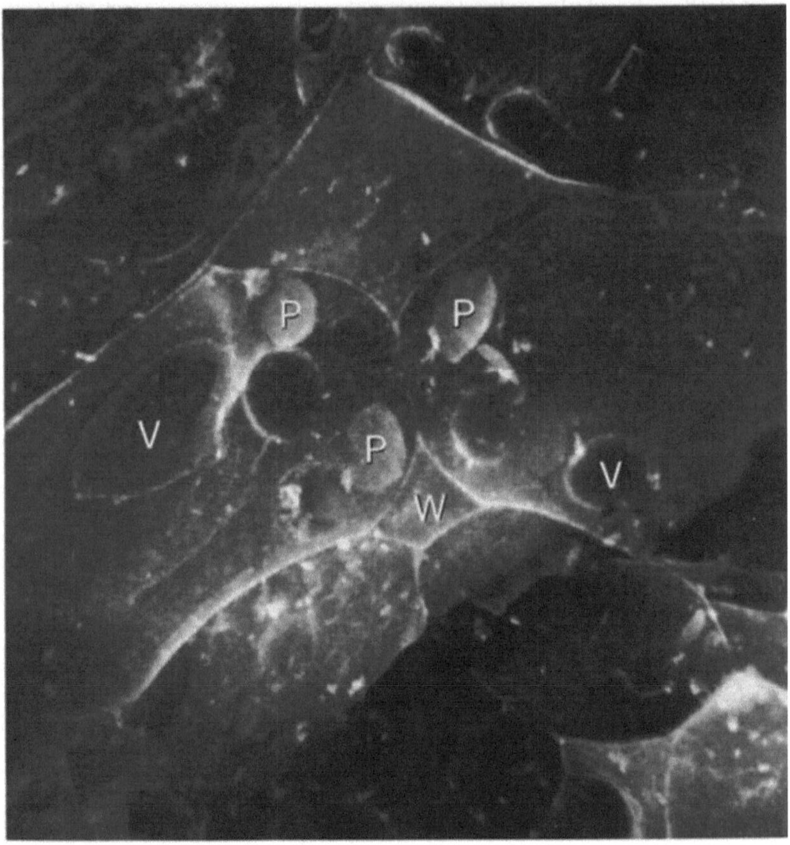

Fig. 1. Frozen-hydrated (unetched) cryofracture surface of *Lemna minor* root in the apical meristem region, encapsulated in 25% PVP. The unetched surface shows little detail although it is possible to identify plastids containing starch grains (*P*), vacuoles (*V*), and a large ice-filled inter-cellular space (*W*). Sample coated with 20 nm of gold at 100 K and examined in the secondary electron emission mode at 20 kV at 120 K on the cold stage of a Cambridge Instruments S-4 SEM. ×5400 (No. 9180)

lessen the penetration of ice into the cell, which MAZUR (1970) considered to be the key step in the freezing process, as such ice crystals nucleate the internal water. The available evidence suggests that they do not exert undue osmotic effects and they are generally less toxic than their penetrating counterparts. Because they do not enter the cells they could be used in connection with freeze-drying procedures. It would also appear useful to investigate the simultaneous use of penetrating and non-penetrating agents. The aim of all such studies should be to measure the success or failure of the various agents in terms of optimal morphologic detail, preservation of vitality, and preven-

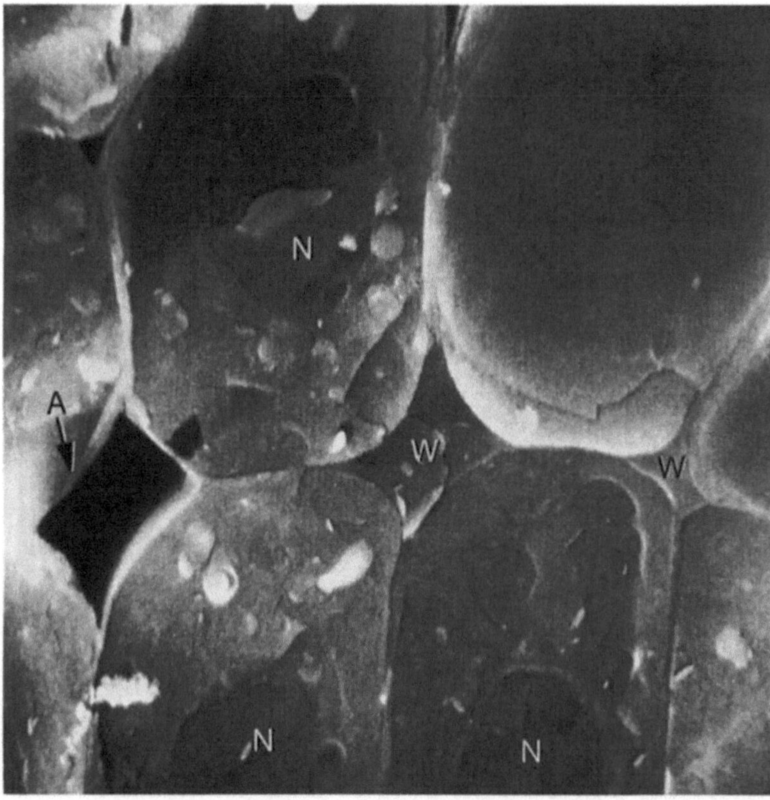

Fig. 2. Partially frozen-hydrated (etched) cryofracture surface *Zea mais* root in the apical meristem region, encapsulated in 25% PVP. There is very little detail in the fracture face other than the faint outline of nuclei (*N*), and possibly some vacuoles and cytoplasmic membranes. The important feature of this picture is the presence of ice-filled (*W*) and air-filled (*A*) inter-cellular spaces. Sample coated with 20 nm of gold at 100 K and examined in the secondary electron emission mode at 20 kV at 120 K on the cold stage of a Cambridge Instruments S-4 SEM. ×1100 (No. 9238)

tion of loss of soluble constituents. While it is hoped that the ideal cryoprotectant will satisfy all three criteria there is no a priori reason why this should be so. The very fact that cells and tissues can survive a particular regime of cryoprotection, freezing, and thawing gives little information that the morphology is unchanged in the frozen state. It would be easier, as REBHUN (1972) suggests, to separate morphologic integrity from the ability to survive and proceed along independent lines in the development of cryobiological techniques. This, however, begs the issue as far as developing adequate techniques to contain the soluble constituents of the cell in their natural loca-

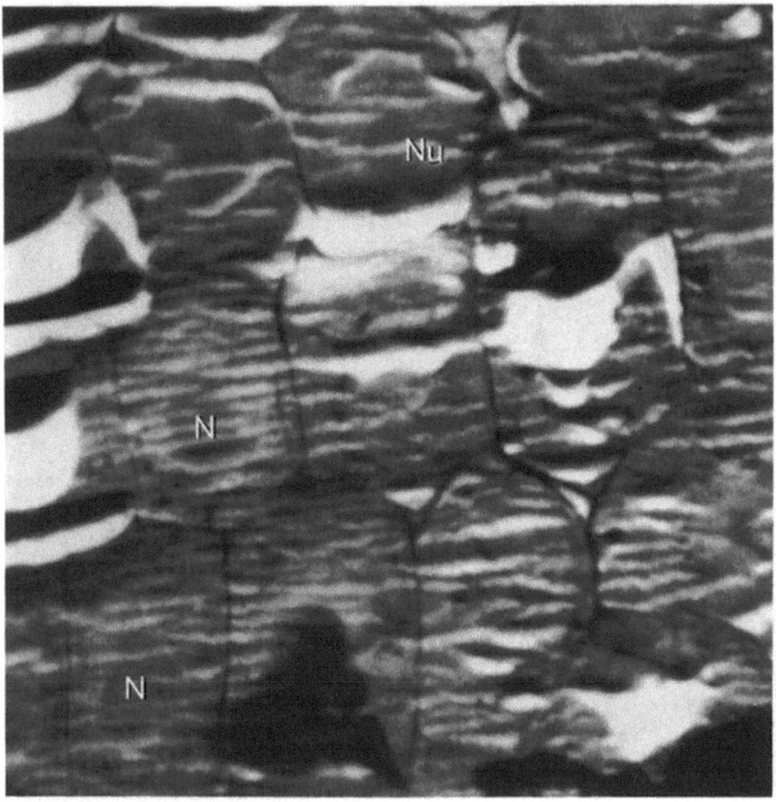

Fig. 3. A 1 – 2 μm frozen hydrated section of *Pisum sativum* root in the apical meristem region. It is possible to identify nuclei (*N*) and nucleoli (*Nu*), although details of the cytoplasm sub-structure are missing. The tissue sections have been used for X-ray microanalysis. Sections cut at 193 K and examined uncoated in the scanning transmission mode at 50 kV at 140 K on the cold stage of JXA-50 A electron probe microanalyser. The lines across the section are due to the cutting, which is considered to be a continuous fracture process rather than a smooth cutting action. ×3000 (No. 6520)

tion. For the time being we have to leave unanswered the question whether the optimal retention of the soluble contents of cells will be best achieved using methods that result in good morphology in the frozen state or using methods that yield highest survival rates after freezing and thawing.

IV. Embedding Agents

BERNARD and VIRON (1971) found that it was useful to embed large pieces of tissue in 10 – 20% gelatin prior to freezing and sectioning in the fro-

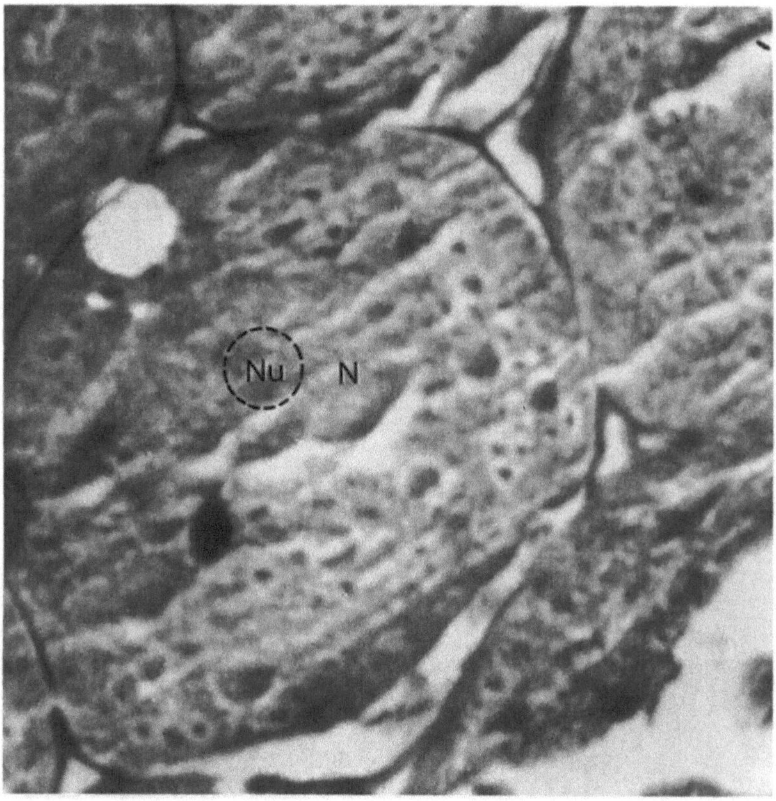

Fig. 4. A 1 – 2 μm frozen-hydrated section of *Pisum sativum* root in the cortical cell region. The nucleus (*N*) and nucleolus (*Nu*) can be seen, although details of the cytoplasm sub-structure are not apparent. Sections cut at 193 K and examined uncoated in the scanning transmission mode at 50 kV at 140 K on the cold stage of a JXA-50 A electron probe microanalyser. ×2100 (No. 6513)

zen state. Experience in this laboratory has indicated that animal tissues rarely require embedding prior to freezing and sectioning but that this procedure is the rule rather than the exception as far as plant tissues are concerned. Finely chopped liver is a good embedding media for root material, but it is necessary to paint the roots with a mixture of olive oil and tetradecane (SANDERSON, 1972) before placing them in the liver. This is to prevent movement of soluble substances between the embedding agent and the specimen. The same precautions are necessary if methyl cellulose is used. We have recently (ECHLIN and MORETON, 1976; and ECHLIN and BURGESS, 1977) been having success with PVP as an embedding agent for *Lemna* roots. It is necessary to make up the PVP in a buffered balanced salts solution; otherwise, the tissues

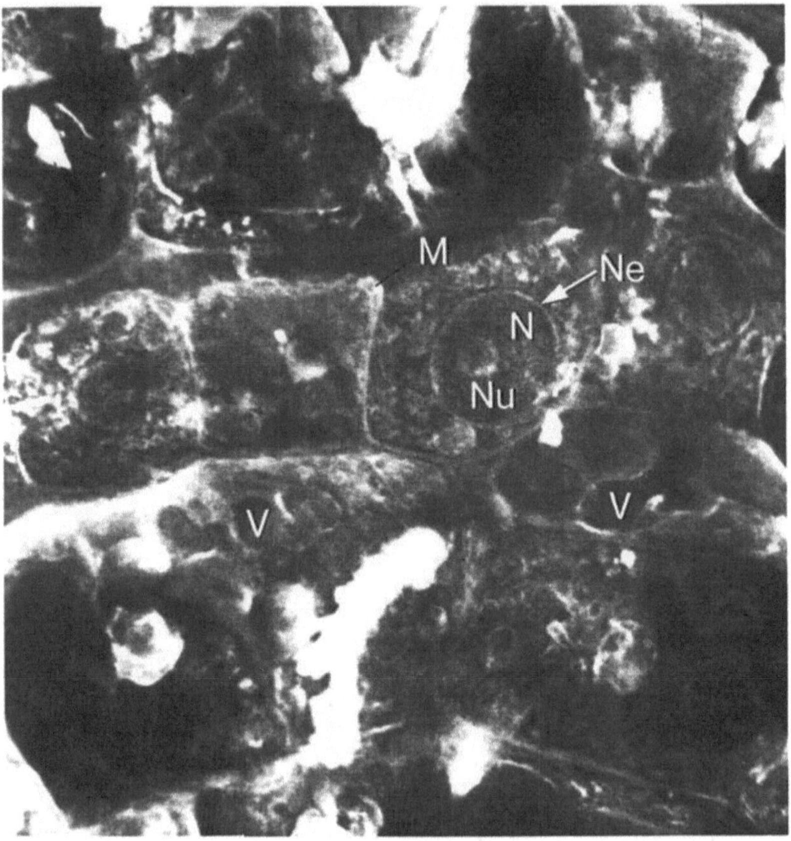

Fig. 5. Partially frozen-hydrated (etched) cryofracture surface of *Lemna minor* root in the apical meristem region encapsulated in 25% PVP. Sublimation of a surface layer of ice reveals considerable detail of the cellular sub-structure. Membranes and cytoplasmic contents may be seen, and it is possible to identify the nucleus (*N*), nucleolus (*Nu*), nuclear envelope (*Ne*), vacuoles (*V*), and the cell membrane (*M*). The large white masses are contaminating ice crystals that are charging in the electron beam. Sample coated with 20 nm of gold at 100 K after fracturing and etching and examined in the secondary electron emission mode at 20 kV at 120 K on the cold stage of a Cambridge Instruments S-4 SEM. ×5400 (No. 9023)

show osmotic damage. The addition of a small amount of graphite to the PVP substantially increases its electrical conductivity and minimizes charging when specimens are examined uncoated in the SEM (Figs. 1 – 7).

The whole purpose of the embedding agent is to provide additional mechanical support for otherwise very brittle tissues that are to be sectioned or

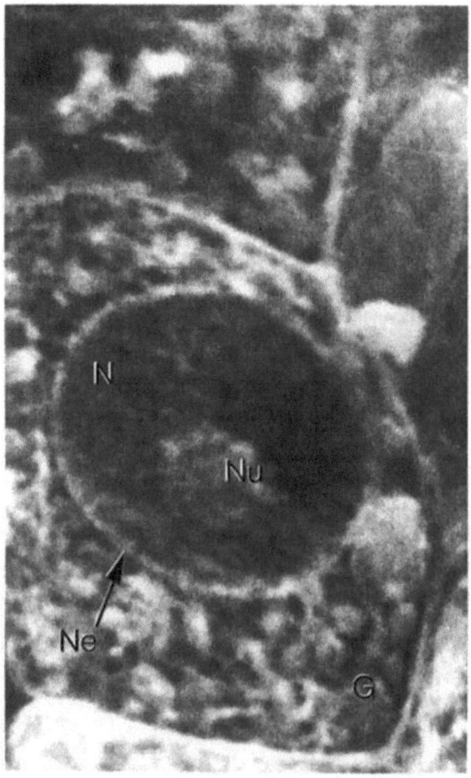

Fig. 6. Partially frozen-hydrated (etched) cryofracture surface of *Lemna minor* in the apical meristem regions of the root, encapsulated in 25% PVP. Detailed view of part of Figure 5 showing the nucleus (*N*), nucleolus (*Nu*), and numerous cytoplasmic granules (*G*), some of which are most certainly mitochondria. ×18,000 (No. 9024)

fractured at low temperatures. Such agents should be non-toxic, cause no undue changes in membrane permeability, and retain a certain amount of flexibility at low temperature. In many ways PVP fulfills most of these objectives.

V. Non-chemical Pre-treatment

The preparative procedures outlined in the preceding sections all suffer from the disadvantage that they involve the addition of foreign substances to the milieu of the cells. It is, however, possible to carry out certain pre-treatment of cells, which as far as is known, is compatible with their life processes. A number of psychrophilic organisms have cryoprotectants as part of their

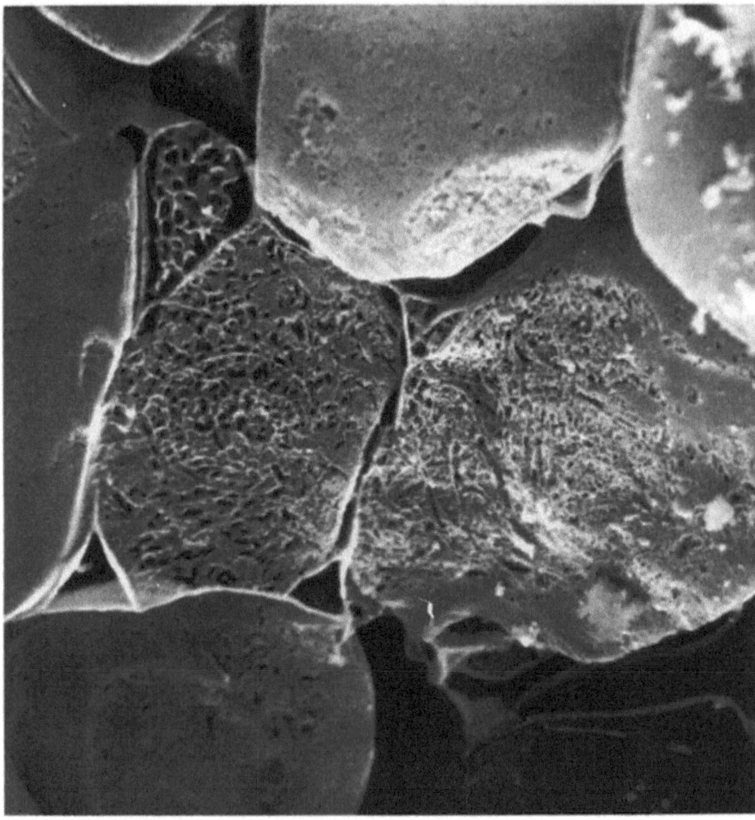

Fig. 7. Frozen-dried cryofractured surface of *Zea mais* root in the apical meristem region, encapsulated in 25% PVP. Although the major cell compartments are visible there is some ice crystal damage in the cytoplasm. All the intercellular spaces appear empty. Material freeze-dried at 200 K at 10^{-5} torr and sputter-coated with 20 nm of gold at room temperature. Examined in the secondary electron emission mode at ambient temperature at 20 kV in a Cambridge Instruments S-4 SEM. ×2000 (No. 9100)

natural body fluids. Immature insects contain a number of different polyhydric alcohols, fish in polar regions contain increased amounts of glycoproteins, and some frost-tolerant plants contain an increased unsaturation of fatty acids and/or an augmentation of membranous material in their cells. Little is known about the mechanisms by which these compounds are formed and the role they play in protecting the cell. BURKE et al. (1976) recently reviewed freezing injury in plants, and their paper contains details of some of the naturally occurring plant cryoprotectants. Attempts should be made to induce frost hardiness in organisms by growing or maintaining them at temperatures below ambient. It has been known for some time that cabbage plants can survive slow freezing (1 – 2 K/h) simply by forming ice in the intra-cellular spaces of the tissue. The plants appear 'wilted' after a heavy frost, but soon return to full turgor as the temperature rises above freezing. Even the simple expedient of cooling specimens to 270 – 275 K before freezing has given promising results. Pure water may be supercooled, and under carefully controlled conditions it is possible to lower the temperature to 238 K before crystallization occurs. However, the presence of any foreign matter promotes heterogeneous nucleation, and it is doubtful whether any significant supercooling could be achieved in biological material, although LYNCH et al. (1965) claim some success with rat liver tissue. KANNOP et al. (1975) have recently shown that it is possible to supercool 'clean' water to 181 K under a pressure of 3 kbars. It is interesting to speculate whether cells and tissues could also survive such pressures and whether such treatment might provide the basis of new methods to freeze biological material. REBHUN (1972) described a series of experiments in which different specimens were coated either with a variety of organic insulating substances or finely powdered nucleating materials before freezing. He found that these surface modification procedures substantially increased cooling rates. MAZUR (1966) found that partial dehydration is a natural phenomenon associated with freeze survival of many cells. Attempts should be made to see how far it is possible to reduce the water content of cells by bathing them in hypertonic solutions before freezing in order to improve their resistance to internal ice formation.

These types of non-chemical pre-treatment warrant more detailed examination as they may well lead to techniques that could result in minimal use of the chemical pre-treatments.

D. Specimen Cooling

The rate of cooling at the centre of a piece of tissue is dependent on the shape and size of the tissue piece, the temperature difference between the tissue and the cooling medium, as well as the density, specific heat, and heat conductivity of both the tissue and the cooling medium. MOOR (1964) showed that the rate at which biological tissues were cooled was the main factor determining the degree of tissue damage in the specimen as measured using the criterion of the preservation of good morphology. A recent paper by the same author (MOOR, 1973) gives an excellent summary of the physical problems associated with quench freezing biological material.

FLETCHER (1971) calculated that a cooling rate of at least 10^{10} K/s would be necessary to vitrify pure bulk water, while RIEHLE (1968 b) calculated that a rate of 10^4 K/s is necessary to vitrify the water in a biological specimen. It is difficult to achieve such rates with reasonably sized biological samples, and the fastest rates that have been achieved so far are between $10^3 - 10^4$ K/s (STEPHENSON, 1960; LUYET et al., 1962; MACKENZIE and LUYET, 1962). If these rates cannot be achieved, then vitrification of the water will not occur and the rapid freezing will favour the formation of a large number of small ice crystals. As the cooling rate becomes progressively slower, so larger crystals are formed in the cell and tissue matrix.

The only practical way at present to achieve rapid cooling is to either rapidly plunge the specimen into a liquid cryogen or to press it against a cooled metal surface. Following the earlier work of ERANKO (1954) and VAN HARREVELD and CROWELL (1964), a number of workers have obtained good preservation in uncryoprotected tissues when they were pressed against a smooth copper block maintained at 77 K. A detailed study by DEMPSEY and BULLIVANT (1976 a, b) confirmed that an ice-free band approximately 12 μm wide can exist in unfixed and uncryoprotected tissue that has been frozen onto a polished copper block at liquid nitrogen temperatures. The copper block method is particularly useful for small specimens, or for parts of larger specimens that are to be sectioned or fractured following the initial freezing. The main advantage of the metal cooling method is that, although a metal such as copper or silver has a thermal capacity about the same as organic liquids, it has a much higher thermal conductivity (405 joule m^{-1} s^{-1} K^{-1} at 173 K) than the liquid cryogens normally used (0.1 joule m^{-1} s^{-1} K^{-1}) and consequently the same amount of heat is transferred 10^4 times faster through copper than through one of the organic cryogens.

However, it is usually more convenient to cool the samples using liquid cryogens. Details relating to the choice and preparation of cryogens are given in an earlier paper (ECHLIN and MORETON, 1973) and in the paper by REBHUN (1972). This latter paper contains a useful discussion of the general principles of freezing in boiling liquids. Although the fluorocarbons are a convenient and relatively efficient way of cooling specimens, it should be remembered that they are potentially carcinogenic, socio-environmentally unacceptable, and can cause contamination of the specimen even at low temperatures.

Although there are a number of hydrocarbons that are useful as quenching agents such as propane, propylene, ethylene, etc., they all suffer the disadvantage that they can form potentially explosive mixtures with oxygen condensed from the surrounding air. They can also dissolve lipid components in the cell, even at quite low temperatures. Many of these disadvantages can be overcome by using a mixture of solid nitrogen and liquid nitrogen. The specific heat of liquid nitrogen maintained at or near its freezing point is sufficiently high to allow very small pieces of tissue to be immersed in it without raising the temperature to its boiling point. Liquid nitrogen slush can be made relatively easily but inefficiently in a vacuum evaporater, or in the specialized piece of equipment described by UMRATH (1974). According to

REBHUN (1972), liquid nitrogen-solid nitrogen slush gives rates up to 10 times faster than that obtained usind liquid nitrogen alone.

Whichever cryogen is used for quench freezing, it is important that the specimen is plunged into it as rapidly as possible. This can be achieved using a simple mechanical injector described by ECHLIN and MORETON (1976) (see Fig. 14) or using the more sophisticated system devised by GLOVER and GARVITCH (1974). Table 1 is a summary of the type of cooling rates that have been achieved by different workers. It can be seen that the cooling rate is between two and seven times faster when the bare thermocouple is mechanically propelled into the cryogen, compared with the rate obtained by hand dipping. When the same thermocouple was immersed into a 1 mm^3 cube of tissue the cooling rate was reduced by a factor of between 5 and 50. Pieces of tissue of these dimensions are the size commonly used in low temperature studies in the SEM, and the cooling rates achieved are closer to those that may be expected using most biological objects.

An acceptable freezing procedure for SEM specimens begins to emerge from the results of these studies together with the immense amount of earlier work on the subject. The tissue samples should be as small as is conveniently possible and placed on very small specimen holders. The sample size must be a compromise between the finitely small samples required for good preservation and the size large enough to handle and examine conveniently and yet avoid the mechanical damage that is invariably found at the surface of specimens excised from a larger piece of tissue. The material may or may not be pre-treated in a number of ways depending on the type of study to be undertaken. The sample should be cooled to as near to its natural freezing point as possible and then rapidly propelled into liquid nitrogen/solid nitrogen slush or pressed against a polished metal surface maintained at liquid nitrogen temperature. After the specimen has been rapidly cooled it should remain under liquid nitrogen to prevent contamination from atmospheric condensation and more important to prevent ice crystal growth.

Once the specimen has been frozen a whole range of specimen manipulations and further treatments can take place. The frozen samples can be subject to analysis and/or morphologic investigation. The low temperature preparative procedures associated with X-ray microanalysis have been the subject of a recent review paper (ECHLIN, 1975 a; ECHLIN and SAUBERMANN, 1977), and it is not proposed to consider this interesting and frequently troublesome aspect of low temperature technology any further. Readers who want further details should refer to the recent books edited by HALL et al. (1974) and ECHLIN and GALLE (1975).

We will now consider the various post-freezing preparative procedures that can be applied to specimens in order to investigate their morphology and ultrastructure.

Table 1. Cooling Rates of Cryogens [K/s 10³] (ECHLIN and MORETON, 1976)

Cryogen	M. P.	Hand rate [a]	Propelled rate [a]	Propelled rates from [b] REBHUN (1972)	Propelled rates from [c] GLOVER (1974)	Propelled rate [b] liver	Propelled rate [b] pea root
Difluorodichloromethane	113 K	–	51.2	3.9	1.8	1.2	0.6
Monofluorodichloromethane	115 K	5.9	43.9	2.9	5.3	1.8	
Liquid nitrogen	77 K	1.0	5.0	–	1.7	1.8	
Iso-pentane (Bottle gas)	121 K	2.9	–	2.4	–		
Monofluorotrichloromethane	92 K	20.8	37.4	1.2	–		
Acetone	178 K	3.9	28.3	–	–		
Liquid nitrogen/solid nitrogen slush	63 K						

[a] Rate from 293 – 223 K, 1 mm × 160 μm thermocouple.
[b] Rate from 273 – 195 K, thermocouple embedded in 2.75 mm stainless steel tube No. 17 gauge.
[c] Rate from 273 – 213 K, 250 μm × 200 μm thermocouple on a 5 mm × 0.1 mm copper disc.

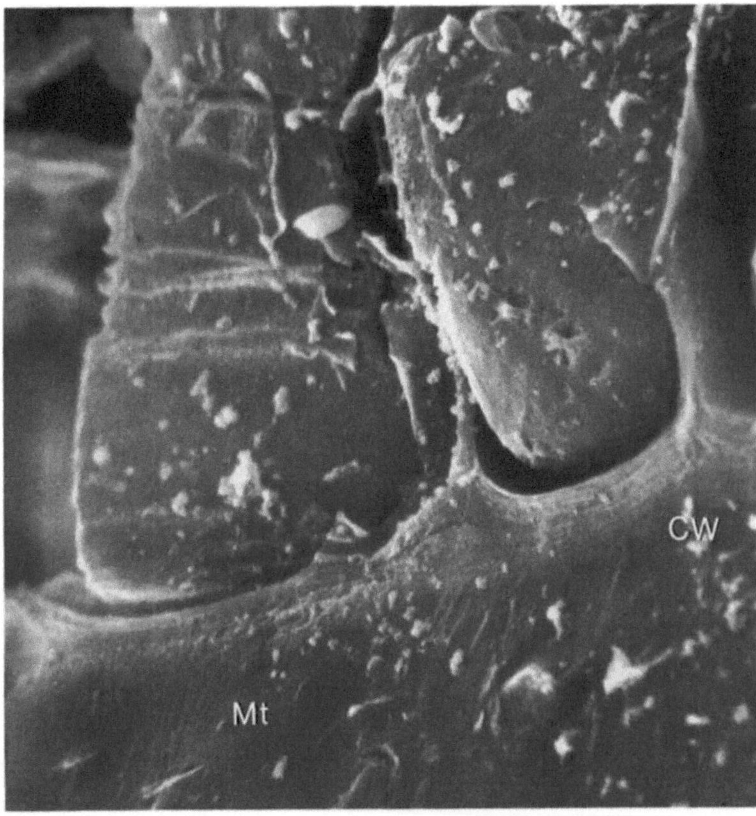

Fig. 8. Frozen-dried cryofractured surface of *Zea mais* root in the apical meristem region, encapsulated in 25% PVP. Details of the cell wall layers (*CW*) can be seen, together with cytoplasmic components (microtubules?) (*Mt*). Specimen shows some shrinkage but very little ice crystal damage. Material freeze-dried at 200 K at 10^{-5} torr and sputter-coated with 20 nm of gold at room temperature. Examined in the secondary electron emission mode at ambient temperature and at 20 kV in a Cambridge Instruments S-4 SEM. ×2000 (No. 9102)

E. Post-freezing Preparative Procedures

There are a number of options open to the experimenter regarding the further treatment of the frozen sample. However, before any treatment is undertaken, it is necessary to decide whether the specimen is to be frozen-dried or frozen-hydrated and whether the external surface or the internal contents are the subject of the investigation.

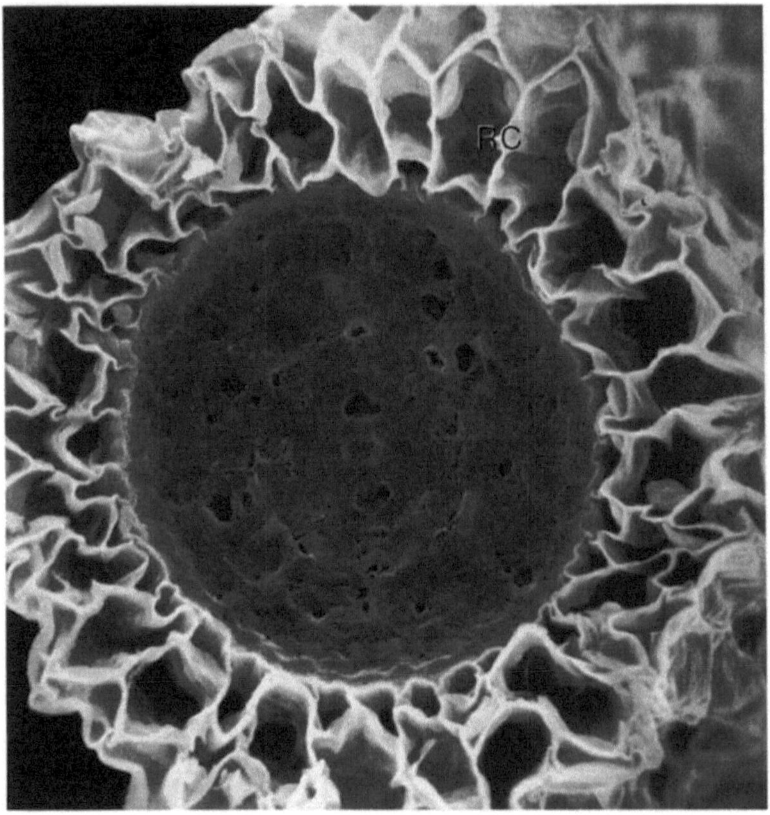

Fig. 9. Ethanol cryofractured surface of *Lemna minor* root in the apical meristem region. The large cells of the root cap (*RC*) can be seen surrounding the symmetrically arranged cells of the central meristematic cylinder. Note the numerous air spaces between the cells. Whole roots fixed in glutaraldehyde, post-fixed in osmium tetroxide, and dehydrated in ethanol. The roots in 100% ethanol were frozen in liquid nitrogen and fractured under the cryogen. Fractured pieces were transferred through several changes of 100% ethanol at room temperature, critical point dried from CO_2, and sputter-coated with 20 nm of gold. Examined in the secondary electron emission mode at ambient temperature and at 20 kV in a Cambridge Instruments S-4 SEM. ×1000 (No. 9518)

I. Frozen-dried or Frozen-hydrated

It is convenient to define these two terms. A frozen-dried specimen is one in which the solid water has been removed either by physical means such as freeze-drying (MERYMAN, 1966) or by chemical means such as freeze-substitution (PEASE, 1973). Both these techniques have been extensively used in

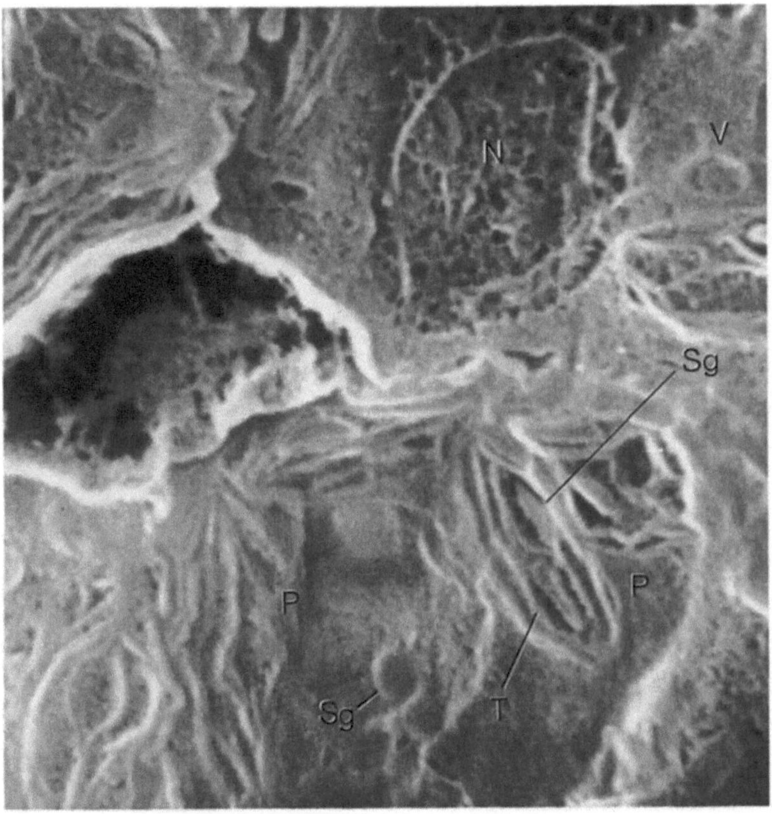

Fig. 10. Ethanol cryofractured surface of *Lemna minor* root in the apical meristematic region. Details of the cytoplasm can be seen including plastids (*P*) with prominent thylakoids (*T*) and starch granules (*Sg*), nuclei (*N*) and vacuoles (*V*). Whole root prepared as for the material shown in Figure 9, and sputter-coated with 20 nm of gold. Examined in the secondary electron emission mode at ambient temperature and at 20 kV in a Cambridge Instruments S-4 SEM. ×10,000 (No. 9771)

the preparation of specimens for SEM and need not be considered further, except to note that it is possible to freeze-dry much faster from a non-polar solvent such as amyl acetate than from water. An example of freeze-dried tissue after encapsulation in 25% PVP is shown in Figure 8. By comparison, a frozen-hydrated specimen is one in which the solid water is retained throughout preparation, examination, and analysis. A recent review (ECHLIN, 1975 a) summarizes much of what has been accomplished using this technique. There are a number of technical difficulties associated with working with frozen-hydrated specimens and not the least of these is the problem of keeping the

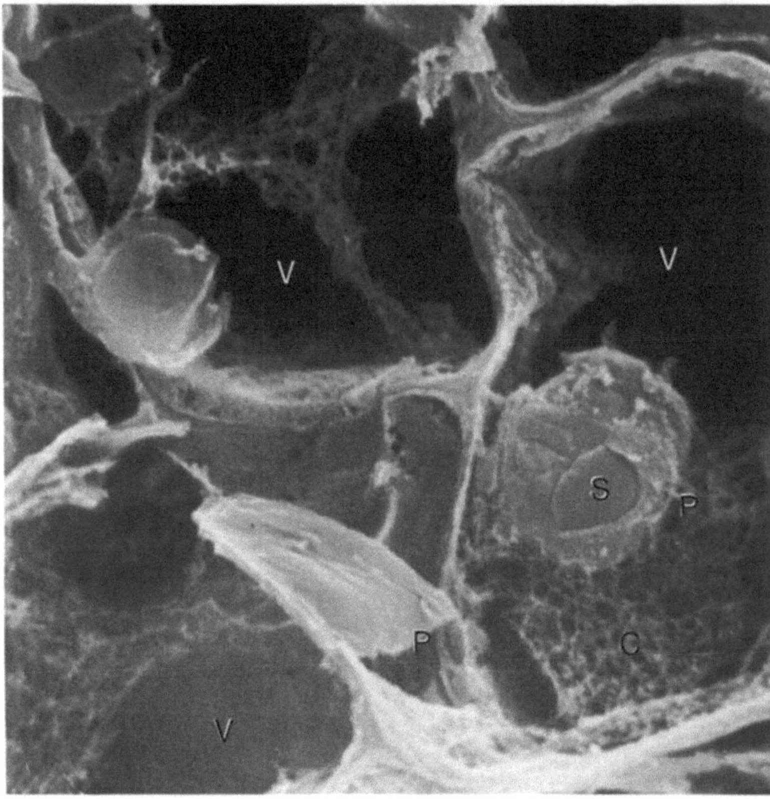

Fig. 11. Resin cryofractured surface of *Lemna minor* root in the outer, partially vacuolate region of the root tip. Details of the plastids (*P*), starch grains (*S*), and peripheral cytoplasm (*C*) can be seen. Much of the cytoplasm has been reduced to a fine fibrillar network and the vacuole (*V*) appears empty. Whole roots fixed in glutaraldehyde, post-fixed in osmium tetroxide, and dehydrated in ethanol. Specimens infiltrated with Spurr's low viscosity resin, frozen in liquid nitrogen, and fractured under the cryogen. Fractured pieces washed several times in 1 : 2 epoxy propane to remove the resin, transferred to ethanol, and critical point dried from CO_2. Specimens were sputter-coated with 20 nm of gold and examined in the secondary electron mode at ambient temperatures at 20 kV in a Cambridge Instruments S-4 SEM. ×5000 (No. 9705)

specimen cold enough inside the SEM to retain the water in the frozen state. Some of these problems are discussed in the recent papers by SAUBERMANN and ECHLIN (1975) and ECHLIN (1978) which also suggests a number of criteria that must be satisfied to ensure that the specimen ist truly frozen-hydrated during examination and analysis in the microscope.

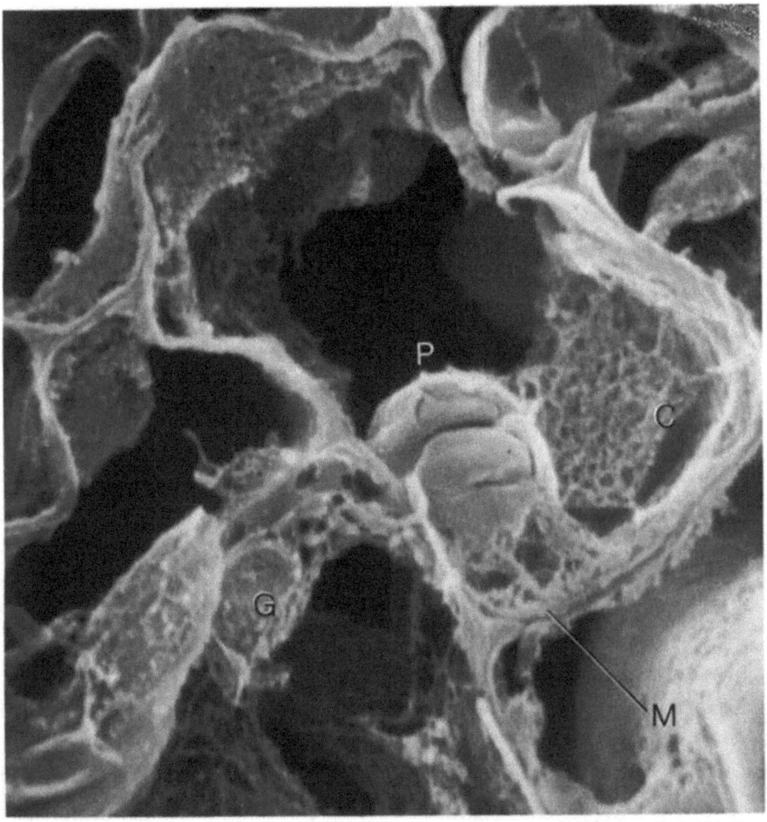

Fig. 12. Resin cryofractured surface of *Lemna minor* root in the outer, partially vacuolate region of the root-tip. The cytoplasm (*C*) appears as a finely fibrillar network in which the plastids (*P*) are embedded. The cell membrane (*M*) can be seen in places as well as a number of granular inclusions (*G*). Whole root prepared as for the material shown in Figure 12, and sputter-coated with 20 nm of gold. Examined in the secondary electron emission mode at ambient temperature and at 20 kV in a Cambridge Instruments S-4 SEM. ×5000 (No. 9706)

II. External Surfaces of Internal Details

A number of papers have already appeared in the literature which demonstrate that it is possible to examine the external features of a wide range of specimens that are maintained in the frozen-hydrated state both coated and uncoated. The uncoated specimens may present special problems with regard to charging, and the instrumental conditions associated with the examination of such specimens will be discussed later.

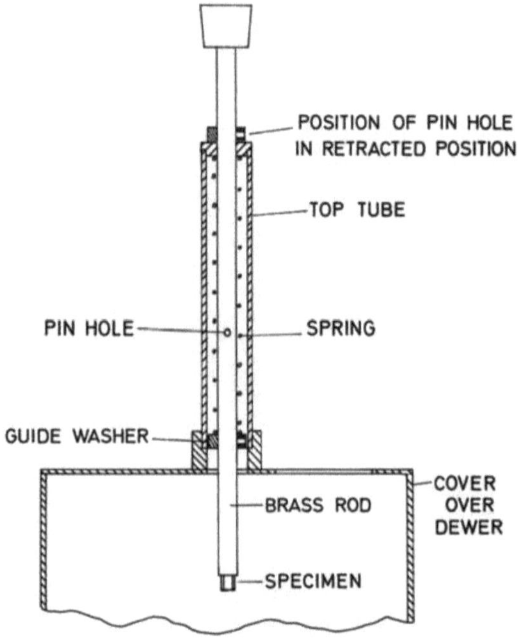

Fig. 13. Specimen-injector mechanism for the rapid propulsion of specimens into liquid cryogens. The specimens, mounted in small silver tubes, are placed at the end of the device and the spring compressed to the retracted position. The injector is positioned over a suitable cryogen and the spring released, which propels the specimen at about 3 m s^{-1} into the cryogen. The silver tube containing the specimen may be released from the injector under the cryogen and the specimen may be sectioned or fractured

It is unfortunate that in many of these investigations on frozen-hydrated specimens scant regard has been paid to ensure that the specimen is indeed frozen-hydrated. A recent paper by CLARK et al. (1976) points out some of the difficulties of accurately measuring the temperature in the region of the specimen. There does, however, seem to be little advantage to be gained in examining the external features of specimens in the frozen-hydrated state. There are serious problems both of contamination and freeze-drying and it would be simpler and more effective to examine surface detail using freeze-dried or critical point dried material. A particularly useful combination of such techniques is to freeze the sample, substitute into alcohol at low temperature, and finally critical point dry from alcohol at ambient temperatures. The real advantage in investigating specimens at low temperatures is in the examination of biphasic biological systems such as lung tissue and some root structures that consist of a mixture of air and water filled spaces. Low tempe-

Fig. 14. Multiple-specimen holder for cryofracturing. The specimens are placed in silver tubes and quench frozen as described in Figure 13. The silver tubes (*S*) are mounted into the recessed holes (*H*) in the holder under liquid nitrogen (see ECHLIN and BURGESS, 1977) and transferred at low temperature to the cold stage of the vacuum evaporator

rature conversion of the aqueous phase to a solid and the retention of this solid in situ enables the experimenter to distinguish between the two types of space. Conventional electron-microscopic preparative techniques including critical point drying and freeze-drying remove the liquid phase, and the end product has a uniform appearance. Studies in progress in this laboratory involve examination of fracture faces of frozen-hydrated *Lemna* roots in an attempt to delineate the small inter-cellular air spaces that appear to spread from the large air-filled lacunae into the meristematic region (Fig. 9, 10).

A number of techniques exist for examining the internal morphology of frozen specimens. All the methods take advantage of the increased brittleness and rigidity of frozen biological material, which permits the fracture or section to be made in the frozen state. Once the specimen has been opened up it may then be examined frozen or the water removed by sublimation or substitution. The simplest way to expose the inside of the specimen is to fracture it crudely under liquid nitrogen. This very effectively exposes an inner surface, but not necessarily the one that is required, and it may be necessary to re-

Fig. 15. Multiple-specimen holder in position on the low temperature stage (*S*) of the vacuum evaporator, with the cryofracturing device (*C*) in position for fracturing. The quartz crystal thin film monitor (*Q*) is used to measure the coating thickness. The heater coil and evaporation source are not shown

peat the process many times before the final answers are obtained. More control over the fracture plane is obtained using the method of HAYES and KOCH (1975), who use a cooled tungsten needle controlled by a micromanipulator, which allow them to carry out real-time dynamic dissection of the specimen. An alternative approach is the one adopted by FUHS and LINDEMANN (1975) who fracture, etch, and subsequently coat the specimen in a freeze etcher and carefully transfer the specimen to the SEM where it is examined and analysed. The logical extension of this approach would be to carry out sequential fracturing in an ante-chamber attached to an SEM, alternately viewing the freshly fractured surface at low temperatures and then exposing a new surface by carrying out another fracture. Once the desired fracture plane has been reached the specimen could be coated at low temperatures (ECHLIN and MORETON, 1973) and viewed at higher resolution in the microscope. As indicated earlier, great care must be taken to ensure that the specimen surface is not contaminated. A review of cryofracturing techniques may be found in a recent paper by ECHLIN and BURGESS (1977), which also gives details of a new way to fracture and coat frozen biological material at low temperatures

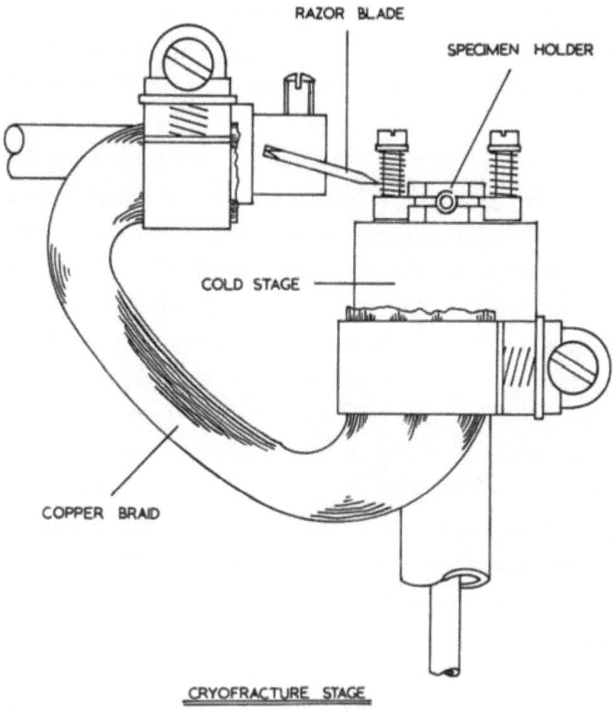

Fig. 16. Diagrammatic view of the cryofracture stage (ECHLIN and BURGESS, 1977) for use at high vacuum in a Balzers vacuum evaporator

in a vacuum evaporator and to examine the etched and unetched material at low temperatures in the SEM (see Figs. 15 – 18).

Another useful approach that has been taken is to fix and dehydrate specimens and to quench freeze from absolute alcohol or from one of the fluorocarbons and then fracture mechanically (HUMPHREY et al., 1974) or non-mechanically (LIM, 1971). Alternatively, the preparative procedure may be continued from dehydration by infiltrating the specimen with paraffin wax or unpolymerized resin, which is then quench frozen and fractured or sectioned at low temperatures (TANAKA, 1974). Following these various organic liquid or resin fracture procedures, the embedding agent is removed and the specimen critical point dried. Examples prepared in this way are shown in Figures 11, 12.

HAGGIS et al. (1976) introduced an interesting variation on these techniques whereby cryoprotectant-infiltrated specimens, after being fractured under liquid nitrogen, are allowed to thaw in buffer before being fixed, dehydrated in ethanol, and critical point dried. The thawing in buffer solubilizes

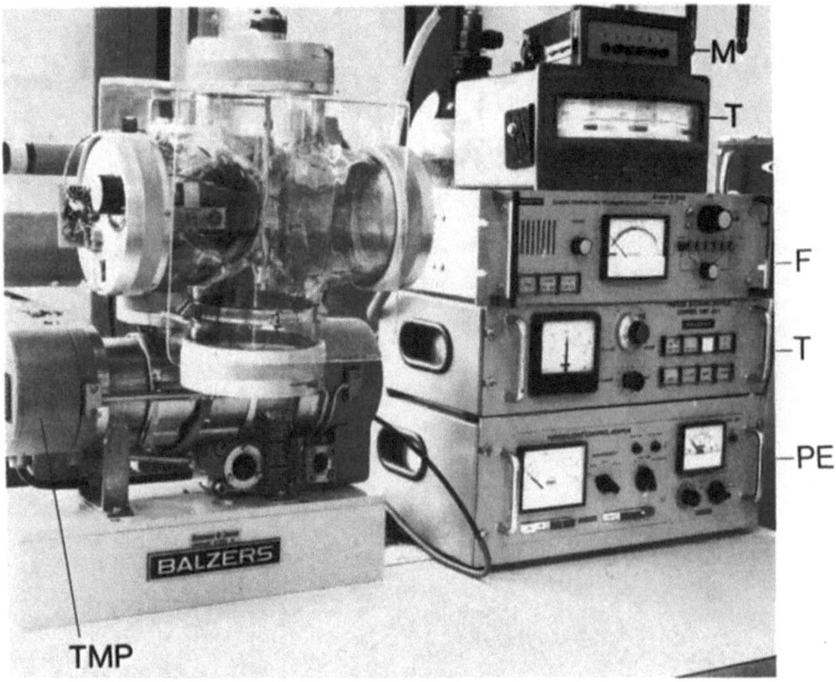

Fig. 17. Balzers vacuum evaporator fitted with a turbo-molecular pump (*TMP*) for use with low-temperature fracturing and metal evaporation techniques. The turbo-molecular pump gives vacuum pressure of about 1×10^{-7} torr with very low contamination levels. Figure 18 shows the relative positions of the cold traps, cold stage, knife assembly, and evaporation sources. A sensitive thermo-galvanometer (*T*) is used to measure the temperature at a number of different places on the specimen by means of a multiple switching device (*M*). The thin film monitor (*F*), temperature control (T_c) and pump and evaporator controls (*PE*) are situated below

and removes some of the low-molecular-weight components, leaving the robust cellular structures more easily visualized.

Finally, the internal contents may be revealed by serial thin (1 – 10 μm) or ultra-thin (100 – 1000 nm) sections that may be cut frozen and examined in the frozen-hydrated or frozen-dried state using the transmission mode of operation of the SEM (MORETON et al., 1974). Although considerable progress has been made in the examination and interpretation of frozen-dried sections, the quality of many fully frozen-hydrated sections continues to leave much to be desired (GUPTA, 1976). It has been suggested (GUPTA et al., 1976) that it would be better to examine partially frozen-hydrated sections that show more recognizable cell morphology in the scanning

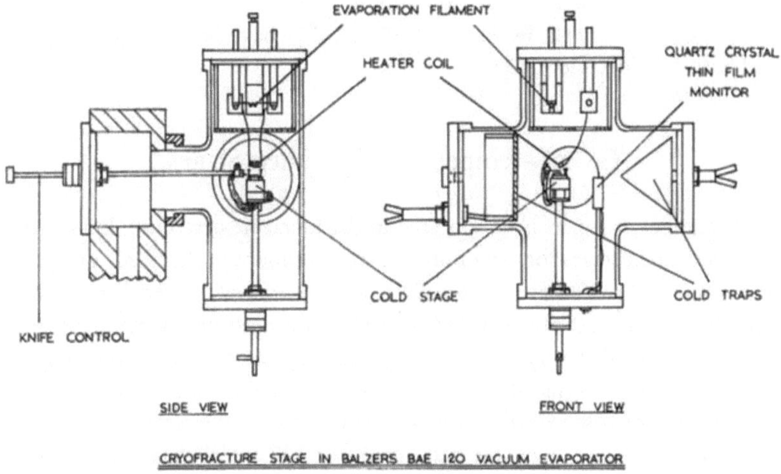

Fig. 18. Diagrammatic view of the cryofracture stage built to operate in the Balzers vacuum evaporator

microscope operating in the transmission mode. However, if the frozen specimens are also to be analysed, such partial drying must be carried out under controlled conditions if accurate local concentration measurements are to be made. An interesting paradox exists as far as analysis and morphology are concerned. The best analytical data appears to be obtained only from fully hydrated material that shows poor morphologic detail. Good ultrastructural detail can only be obtained from either fully dried or partially dried material in which the diffusible substances have either migrated or departed entirely.

F. Specimen Transfer

With very few exceptions, the preparation of material by low temperature methods is carried out in equipment separate from the equipment used for examination and analysis. Care must be taken in transferring specimens, particularly those maintained at low temperatures, and a recent paper (ECHLIN, 1975 a) has reviewed the types of low temperature devices that can facilitate these transfers. Ideally, the specimen should be kept at as low a temperature as possible and in a good vacuum, and a number of devices maintain such conditions. The system described by SAUBERMANN et al. (1977) transports the frozen specimen in a portable cold stage fitted with rotary and diffusion pumps. The specimen may be fractured and coated under high vacuum and transferred to the cold stage of the microscope through a double air-lock system. Yet despite this absolute requirement a number of much simpler devices (see, e. g., SAUBERMANN and ECHLIN, 1975) have been constructed and used with considerable success.

G. Low-temperature Specimen Stages

A number of papers have appeared in the literature describing low-temperature specimen stages for examining frozen specimens in the SEM. The problems associated with stage design have been extensively covered in a recent paper by SAUBERMANN and ECHLIN (1975) and need not be repeated here. Suffice it to say that very few cold stages are capable of maintaining a specimen cold enough so that it may be examined frozen-hydrated in the high vacuum of the microscope. It is not sufficient to record the temperature of the cold stage, for this is anywhere between 30 – 60 K colder than the specimen. The only really efficient cold stages are those that bring liquid nitrogen as close as possible to the specimen and still maintain reasonable mechanical stability. Many of the low-temperature SEM studies described in the literature are probably of frozen-dried or at best partially frozen-hydrated material. The first generation of cold stages that are presently available on most commercial SEM have served their purpose well. We now need more efficient stages that can effectively cool specimens to temperatures at which we can be sure they are fully frozen-hydrated.

H. Specimen Examination

The only real problems associated with the examination of specimens comes with bulk material that is maintained in the frozen-hydrated state on a cold stage inside the microscope. Such specimens may charge and in many cases may begin to freeze-dry in the microscope. This is because the surface of the specimen most distant from the cold stage is the one examined in the microscope. Although the energy input from the electron beam is relatively small (CLARK et al., 1976), there is a large thermal gradient between the top and the bottom of the specimen. This reinforces the argument that the cold stage of an electron beam instrument should be maintained at as low a temperature as possible. The problems are less acute with sections that are examined in either the transmission or secondary mode, provided the sections are flattened and in good thermal contact with a thin conducting support film or metallic grid because most of the energy in the electron beam passes through the specimen. There are also problems of specimen examination in investigations on whole or fractured bulk material that have not been coated with a thin layer of conducting metal or carbon. Under these conditions it is advisable to use the lowest accelerating voltage and beam current conducive to the optimal resolution and information that is required of the specimen. It has been known for some time that there is an increase in signal current at low kV. The production efficiency of secondary and back-scattered electrons is better and specimen penetration is reduced. Charging is reduced (on clean specimens) because the average number of secondary electrons emitted from the material for each incident primary electron is greater than one at low accelerating voltages. It is also useful to use beam-blanking devices and slower raster speeds to decrease the time that the incident beam remains on a single spot.

Although considerable work has been carried out on the conductivity of pure ice (see HOBBS, 1974), very little work has been done on the conductivity of frozen-hydrated biological material. In a series of studies on insect tissue and gels containing known amounts of electrolytes, MARSHALL (1975) showed that there was diminished charging in frozen-hydrated bulk specimens that had been surface-dehydrated inside the microscope. Fully frozen-hydrated specimens charged even at low kV, although little or no charging was experienced in specimens coated with carbon. Marshall considers that the charging is due to a decrease in charge carrier mobility at low temperatures (123 – 133 K). The same type of problem has been considered in some detail by FUHS and LINDEMANN (1975) and BROMBACH (1975). They found that because ice is such a good insulator at 128 K a space charge quickly builds up in the sample, which in turn modifies the dimensions of the electron beam/specimen interaction volume and may diminish the intensity of the high energy end of the X-ray spectrum. Such phenomena are alleviated following the application of a thin layer of evaporated carbon. It should be noted that these studies were carried out at relatively low kV on pure amorphous ice specimens, and it remains to be seen whether the same phenomena exist in bulk frozen-hydrated biological material. A more encouraging picture emerges from the work of WITTRY and WU (1975) who present a model that explains why frozen biological specimens can be examined in the SEM. They show that there is a stable operating point determined by the interrelationship between specimen current, beam landing energy, and the effective resistance of the specimen. They show that the specimen current determines surface charging voltage, and that at low voltages specimen resistivity is lower than that predicted by theory. In contrast to MARSHALL (1975), WITTRY and WU find that there is an increase in the number of carriers that are produced by soft X-ray characteristics and recombination radiation.

I. Conclusions

An attempt has been made to review the advantages and disadvantages of the low temperature techniques that are presently available for the preparation, examination, and analysis of biological material in the SEM. Some of the disadvantages are quite severe, and they centre around the problems associated with transferring the water in the specimen from the liquid to the solid state. Once this state has been achieved, however, it is relatively easy to carry out a number of manipulations on the specimen. There are a number of extra advantages that accrue from working at low temperatures in electron beam instruments. The contamination rate is much lower (ECHLIN, 1975 b) and there is less thermal damage (HALL and GUPTA, 1974) and radiation damage (TAYLOR and GLAESER, 1976).

References

Bachmann, L., Schmitt, W. W.: Improved cryofixation applicable to freeze-etching. Proc. Natl. Acad. Sci. U. S. **68**, 2149 (1971)

Bernhard, W., Viron, A.: Improved techniques for the preparation of ultra-thin frozen sections. J. Cell Biol. **49**, 731 – 746 (1971)

Boyde, A.: Freezing, freeze fracturing and freeze-drying in biological specimen preparation for the SEM. In: Proc. 7th Ann. IITRI-SEM Symp., Chicago, Ill.: Chicago Press Corp., 1974, pp. 1043 – 1046

Boyde, A., Echlin, P.: Freezing and freeze-drying: a preparative technique for SEM. In: Proc. 6th Ann. IITRI-SEM Symp., Chicago, Ill.: Chicago Press Corp., 1973, pp. 759 – 766

Brombach, J. D.: Electron beam X-ray microanalysis of frozen biological bulk specimens below 130 K. II. The electrical charging of the sample in quantitative analysis. J. Micros. Biol. Cell. **22**, 233 – 238 (1975)

Burke, M. J., Gusta, L. V., Quamme, H. A., Weiser, C. J., Li, P. H.: Freezing and injury in plants. Ann. Rev. Plant Physiol. **27**, 507 – 528 (1976)

Clark, J., Echlin, P., Moreton, R. B., Saubermann, A. J., Taylor, P. J.: Thin-film thermocouples for SEM. In: Proc. 9th Ann. IITRI-SEM Symp. Chicago, Ill.: Chicago Press Corp., 1976, pp. 83 – 90

Dempsey, G. P., Bullivant, S.: A copper block method for freezing non-cryoprotected tissue to produce ice-crystal free regions for electron microscopy. I. Evaluation using freeze-substitution. J. Microscopy **106**, 251 – 260 (1976 a)

Dempsey, G. P., Bullivant, S.: A copper block method for freezing non-cryoprotected tissue to produce ice-crystal free regions for electron microscopy. II. Evaluation using freeze-fracturing with a cyro-ultramicrotome. J. Microscopy **106**, 261 – 271 (1976 b)

Dorge, A., Rick, R., Gehring, K., Mason, J., Thurau, K.: Preparation and applicability of freeze-dried sections in the microprobe analysis of biological soft tissue. In: Biological Microanalysis. Echlin, P., Galle, P. (eds.). Paris: Soc. Franc. Micros. Électronique 1975

Echlin, P.: Scanning microscopy at low temperature. In: Freeze-etching: Techniques and applications. Benedetti, E. L., Favard, P. (eds.). Paris: Soc. Franç. Micros. Électronique. 1973, Chap. 17

Echlin, P.: The preparation of frozen-hydrated biological material for X-ray microanalysis. J. Micros. Biol. Cell. **22**, 215 – 226 (1975 a)

Echlin, P.: Contamination in the scanning electron microscope. In: Proc. 8th Ann. IITRI-SEM Symp., Chicago, Ill.: Chicago Press Corp., 1975 b, pp. 679 – 686

Echlin, P.: Low temperature scanning electron microscopy. J. Micros. **112** (in press) 1978

Echlin, P., Burgess, T.: Cryofracturing and low temperature scanning electron microscopy of plant material. In: Proc. 10th Ann. IITRI-SEM Symp. Chicago, Ill.: Chicago Press Corp., 1977, pp. 491 – 500

Echlin, P., Galle, P. (eds.): Biological Microanalysis. Paris: Soc. Franç. Micros. Électronique 1975

Echlin, P., Moreton, R. B.: The preparation, coating and examination of frozen biological materials in the SEM. In: Proc. 6th Ann. IITRI-SEM Symp., Chicago, Ill.: Chicago Press Corp., 1973, pp. 325 – 332

Echlin, P., Moreton, R. B.: Low temperature techniques for scanning electron microscopy. In: Proc. 9.th Ann. IITRI-SEM Symp. Chicago, Ill.: Chicago Press Corp., 1977, pp. 753 – 762

Echlin, P., Saubermann, A. J.: Preparation of biological specimens for X-ray microanalysis. In: Proc. 10th Ann. IITRI-SEM Symp. Chicago, Ill.: Chicago Press Corp., 1977, pp. 621 – 638

Echlin, P., Skaer, H. le B., Gardiner, B. O. C., Franks, F., Asquith, M. H.: Polymeric cryoprotectants in the preservation of biological ultrastructure. II. Physiological effects. J. Micros. **110** (3) (in press) 1977

Eranko, O.: Quenching of tissues for freeze-drying. Acta anat. (Basel) **22**, 331 – 336 (1954)

Fletcher, N. H.: Structural aspects of the ice-water system. Rep. Prog. Phys. **34**, 913 – 993 (1971)

Franks, F., Asquith, M. H., Hammond, C. C., Skare, H. le B., Echlin, P.: Polymeric cryoprotectants in the preservation of biological ultrastructure. I. Low temperature states of aqueous solutions of hydrophilic polymers. J. Micros. **110** (3) (in press) 1977

Franks, F., Skaer, H. Le B.: Aqueous glasses as matrices in freeze fracture electron microscopy. Nature (London) **262**, 323 – 325 (1976)

Fuhs, W., Lindemann, B.: Electron beam X-ray microanalysis of frozen bulk specimens below 130 K. I. Instrumentation and specimen preparation. J. Micros. Biol. Cell. **22**, 227 – 232 (1975)

Glover, A. J., Garvitch, Z. S.: The freezing rate of freeze-etch specimens for electron microscopy. Cryobiology **11**, 248 – 254 (1974)

Gupta, B. L.: Water movement in cells and tissues. In: Perspectives in Experimental Biology. Spencer Davies, P. (ed.). Oxford: Pergamon Press, 1976, Vol. I, pp. 25 – 42

Gupta, B. L., Hall, T. A., Madrell, S. H. P., Moreton, R. B.: The distribution of ions in fluid transporting epithelium determined by electron probe microanalysis. Nature **264**, 284 – 286 (1976)

Haggis, G. H., Bond, E. F., Phipps, B.: Visualization of mitochondrial cristae and nuclear chromatin by SEM. In: Proc. 9th Ann. IITRI-SEM Symp., Chicago, Ill.: Chicago Press Corp., 1976, pp. 281 – 286

Hall, T. A., Echlin, P., Kaufmann, R. (eds.): Microprobe Analysis as Applied to Cells and Tissues. London: Academic Press 1974

Hall, T. A., Gupta, B. L.: Beam induced loss of organic mass under electron microprobe conditions. J. Microscopy **100,** 177 – 188 (1974)

Hayes, T. L., Koch, G.: Some problems associated with low temperature micromanipulation in the SEM. In: Poc. 8th Ann. IITRI-SEM Symp. Chicago, Ill.: 1975, pp. 35 – 42

Heide, H. G. Grund, S. J.: Eine Tiefkühlkette zum Überführen von wasserhaltigen biologischen Objekten ins Elektronenmikroskop. J. Ultrastruct. Res. **48,** 259 – 268 (1974)

Hobbs, P. V.: Ice Physics. Oxford: Clarendon Press 1974

Holbrook, K. A., Holbrook, J. R., Odland, G. F.: A comparison of the effects of sample preparation on the X-ray microanalysis and ultrastructure of the thin sectioned specimens of human skin. In: Proc. 9th Ann. IITRI-SEM Symp. Chicago, Ill.: Chicago Press Corp., 1976, pp. 266 – 274

Humphrey, W. J., Spurlock, B. O., Johnson, J. S.: Critical point drying of ethanol-infiltrated cryofractured biological specimens for SEM. In: Proc. 7th Ann. IITRI-SEM Symp., Chicago, Ill.: Chicago Press Corp., 1974, pp. 275 – 282

Kanno, H., Speedy, R. J., Angell, C. A.: Supercooling of water to –92° C under pressure. Science **189,** 880 – 881 (1975)

Lim, D. J.: SEM observations on non-mechanically cryofractured biological tissue. In: Proc. 4th Ann. IITRI-SEM Symp. Chicago, Ill.: Chicago Press Corp., 1971, pp. 257 – 264

Luyet, B. J., Tanner, J., Rapatz, G.: X-ray diffraction study of the structure of rapidly frozen gelatin solutions. Biodynamica **9,** 21 – 46 (1962)

Lynch, R., Bitensky, L., Chayen, J.: On the possibility of supercooling in tissues. J. Roy. Micros. Soc. **85,** 213 – 222 (1965)

MacKenzie, A. P.: Freezing, freeze-drying and freeze-substitution. In: Proc. 5th Ann. IITRI-SEM Symp., Chicago, Ill.: Chicago Press Corp., 1972, pp. 273 – 280

MacKenzie, A. P., Luyet, B. J.: A colloidin sandwich-film technique for the study of growth of ice in very thin layers of aqueous solutions. In: Proc. 5th Int. E. M. Meeting. Philadelphia: Academic Press 1962, Vol. II, p. 2

Marshall, A. T.: X-ray microanalysis of frozen-hydrated biological specimens: The effect of charging. Micron **5,** 275 – 280 (1975)

Mazur, P.: Physical and chemical basis of injury in single-celled microorganisms subjected to freezing and thawing. In: Cryobiology. Meryman, H. T. (ed.). New York: Academic Press 1966

Mazur, P.: Cryobiology: The freezing of biological systems. Science **168,** 939 – 946 (1970)

Meryman, H. T.: Freeze drying. In: Cryobiology. Meryman, H. T. (ed.). New York: Academic Press 1966

Meryman, H. T.: Freezing injury and its prevention in living cells. Ann. Rev. Biophys. Bioengineering **3,** 341 – 363 (1974)

Moor, H.: Die Gefrierfixation lebender Zellen und ihre Anwendung in der Elektronenmikroskopie. Z. Zellforsch. **62,** 546 – 589 (1964)

Moor, H.: Cryotechnology for the structural analysis of biological material. In: Freeze etching: techniques and applications. Benedetti, E. L., Favard, P. (eds.). Paris: Soc. Franç. Micros. Électronique 1973

Moreton, R. B., Echlin, P., Gupta, B. L., Hall, T. A., Weis-Fogh, T.: Preparation of frozen-hydrated tissue sections for X-ray microanalysis. Nature (London) **247**, 113 – 115 (1974)

Pease, D. C.: Substitution techniques. In: Advanced Techniques in Biological Electron Microscopy. Koehler, J. K. (ed.). Berlin: Springer 1973

Plattner, H., Fisher, W. M., Schmitt, W. W., Bachmann, L.: Freeze etching of cells without cryoprotectives. J. Cell Biol. **53**, 116 – 126 (1972)

Rebhun, L. I.: Freeze-substitution and freeze-drying. In: Principles and Techniques of Electron Microscopy. Biological Applications. Hayat, M. A. (ed.). New York: Van Nostrand, 1972, Vol. 2, pp. 3 – 42

Riehle, U.: Über die Vitrifizierung verdünnter wässriger Lösungen. Diss. No. 4271 ETH Zürich: Juris Verlag 1968 a

Riehle, U.: Schnellgefrieren organischer Präparate für die Elektronenmikroskopie. Chemie-Ingenieur-Technik **40**, 213 – 218 (1968 b)

Robards, A. W.: Ultrastructural methods of looking at frozen cells. Sci. Prog. (Oxford) **61**, 1 – 40 (1974)

Sanderson, J.: Microautoradiography of diffusible ions in plant tissues: problems and methods. J. Microscopy **96**, 245 – 254 (1972)

Saubermann, A. J., Echlin, P.: The preparation, examination and analysis of frozen-hydrated tissue sections by SEM and X-ray microanalysis. J. Microscopy **105**, 155 – 192 (1975)

Saubermann, A. J., Riley, W., Echlin, P.: Preparation of Frozen Hydrated Tissue Sections for X-ray microanalysis using a satellite vacuum coating and transfer system. Proc. 10th Ann. IITRI-SEM Symp., Chicago, Ill.: Chicago Press Corp., 1977, pp. 347 – 356

Sjostrom, M., Thornell, L. E.: Preparing sections of skeletal muscle for transmission electron analytical microscopy (TEAM) of diffusible elements. J. Microscopy **103**, 101 (1976)

Skaer, H. le B., Franks, F., Asquith, M. H., Echlin, P.: Polymeric cryoprotectants in the preservation of biological ultrastructure. III. Morphological aspects. J. Micros. **110** (3) (in press) 1977

Stephenson, J. L.: Fundamental problems in freezing and drying of biological materials. In: Recent Researches in Freezing and Drying. Parkes, A. S., Smith, A. V. (eds.). Oxford: Blackwell, 1960

Tanaka, K.: Frozen resin cracking method and its role in cytology. In: Principles and Techniques of SEM. Hayat, M. A. (ed.). New York: Van Nostrand, 1974, Vol. I, Chap. 4

Taylor, K. A.: Electron microscopy of frozen, hydrated biological specimens. In: Proc. 33rd Ann. Meeting EMSA, Bailey, G. W. (ed.). Baton Rouge, LA: Claitor's Publishing Division, 1975, pp. 300 – 301

Taylor, K. A., Glaeser, R. M.: A method for maintaining specimen hydration by sandwiching between thin films. In: Proc. 31st Ann. Meeting EMSA, Arceneaux, C. J. (ed.). Baton Rouge, LA: Claitor's Publishing Division, 1973, pp. 342 – 343

Taylor, K. A., Glaeser, R. M.: Electron microscopy of frozen hydrated biological specimens. J. Ultrastruct. Res. **55**, 448 – 456 (1976)

Umrath, W.: Cooling bath for rapid freezing in electron microscopy. J. Microscopy **101**, 103 – 106 (1974)

Van Harreveld, A., Crowell, J.: Electron microscopy after rapid freezing on a metal surface and substitution fixation. Anat. Rec. **149**, 381 (1964)

Wittry, D. B., Wu, C. J.: The charging of semi-insulating specimens in electron microprobe instruments. In: Proc. 8th Ann. IITRI-SEM Symp., Chicago, Ill.: Chicago Press Corp., 1975, pp. 441 – 446

Quantitative Electron Microscopy of Nucleic Acids

J. FERGUSON and R. W. DAVIS

A. Introduction

This Chapter is not intended to be a comprehensive literature survey, but rather an introduction to techniques that are now available for preparing nucleic acid samples for electron microscopy. In particular, an effort has been made to emphasize a quantitative approach to the electron microscopy of nucleic acids and to give examples of quantitative applications of the techniques discussed. We have often used examples from our own research to demonstrate ways in which quantitative electron microscopy can be done; however, this does not imply that these are the only examples available. YOUNGHUSBAND and INMAN (1974) published a review of the electron microscopy of DNA that contains extensive literature references, and CHOW et al. (1975) prepared a laboratory manual for the electron microscopy of nucleic acids containing extensive information about techniques, instruments, and supplies.

B. Basic Protein Film Method

The basic protein film technique, developed by KLEINSCHMIDT et al. (KLEINSCHMIDT and ZAHN, 1959; KLEINSCHMIDT, 1968), allows visualization in the electron microscope of nucleic acid molecules by virtue of a collapsed column of basic protein around the DNA or RNA. This technique, because of its simplicity and reproducibility, has provided a cornerstone for the development of quantitative electron microscopy of nucleic acids. Subsequent modifications in the technique, such as the use of formamide to destabilize random base interactions and extend single-stranded DNA (WESTMORELAND et al., 1969; DAVIS et al., 1971), have increased its range of applications.

We will describe in detail both the aqueous and formamide variations of the basic protein film technique, and then make some general comments about reagents and problems related to contrast. In the aqueous technique, double-stranded DNA appears as a gently curved filament, and single-stranded DNA as condensed bushes because of random base interactions. This

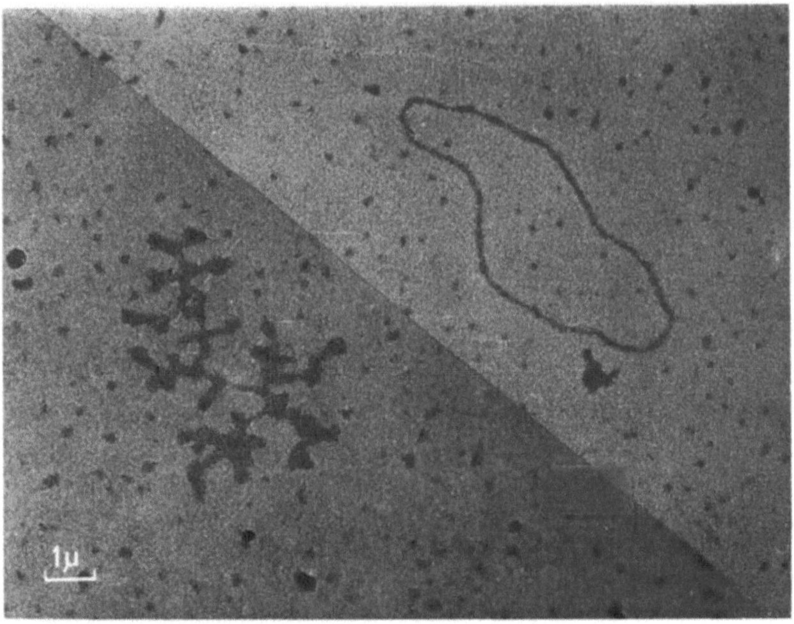

Fig. 1. DNA prepared by the aqueous basic protein film technique. Double-stranded DNA (ΦX174 RFII) appears extended, and single-stranded DNA (λ) appears as a collapsed bush because of random base interactions. The DNA was stained with uranyl acetate

technique is useful for measuring double-strand lengths, measuring the positions of single-strand regions along the double strand, distinguishing positively between single- and double-stranded DNA (Fig. 1), and identifying closed circular molecules with superhelical turns (Fig. 2). In the formamide technique, the DNA is mounted at a formamide and salt concentration such that double-stranded DNA is stable, but the random base interactions in single-stranded DNA are melted out and single-stranded DNA also appears as a curved filament that is distinguishable from double-stranded DNA by being thinner and somewhat more kinky (Fig. 3). Other denaturing agents besides formamide have been tried, although with less success (DAVIS et al., 1971). This technique is useful for measuring single-strand lengths and for observing the topology at the connections between single- and double-stranded regions.

I. Aqueous Technique

The concentrations in the spreading solution are as follows: 0.5 μg/ml of DNA, 0.1 mg/ml of cytochrome c, 0.5 M ammonium acetate (pH 7 to 8),

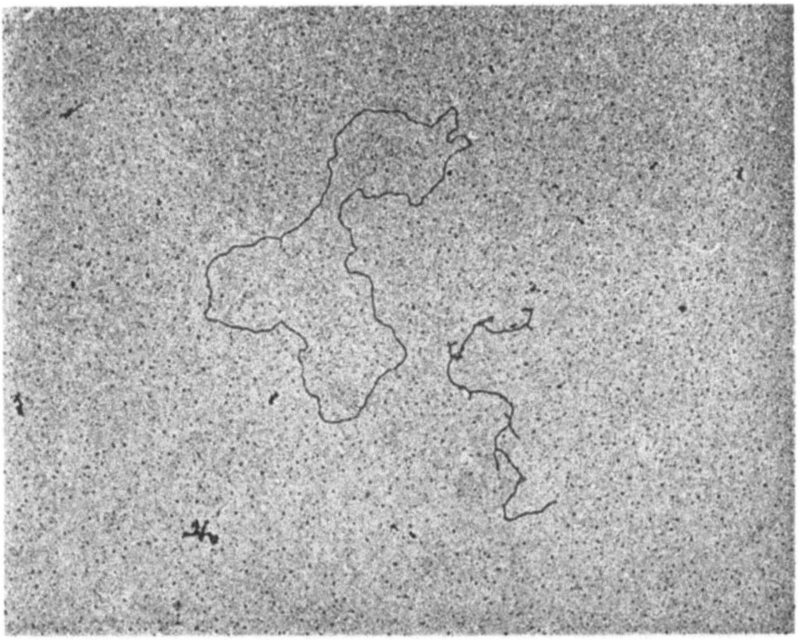

Fig. 2. DNA prepared by the aqueous basic protein film technique. The open circular molecule appears relaxed and the closed circular molecule contains tightly twisted superhelical turns. The DNA is colEl-Trp and is stained with uranyl acetate

and 1 mM Na_3EDTA. Na_3EDTA solutions should be made by NaOH titration from Na_2EDTA, and not from the acid form. The acid form frequently contains impurities that prevent the film from forming. The hypophase, 0.25 M ammonium acetate (pH 7.5), is in a 90-mm square Teflon dish or plastic petri dish. A little talc can be dusted on the hypophase to visualize and compress the film. The clean glass slide serving as a ramp should be rinsed with 0.25 M NH_4Ac and allowed to dry immediately before applying the spreading solution to ensure that it is as clean and dust-free as possible. Glass slides are cleaned in chromic acid, rinsed extensively with water, and stored in 0.25 M NH_4Ac in the absence of Na_3EDTA. A polyethylene tubing-Hamilton syringe combination is used to deliver 30 μl of spreading solution to the slide. The solution is applied slowly with a back-and-forth motion across the slide, about 1 cm above the surface of the hypophase. The solution runs down the slide and the protein film spreads, pushing the talc away. For best results the film is picked up at once onto the grid at a distance of 1 grid width from the slide-solution boundary. The grid is dipped for 30 s into the uranyl stain solution, then rinsed for 10 s in 90% ethanol. The stain solution,

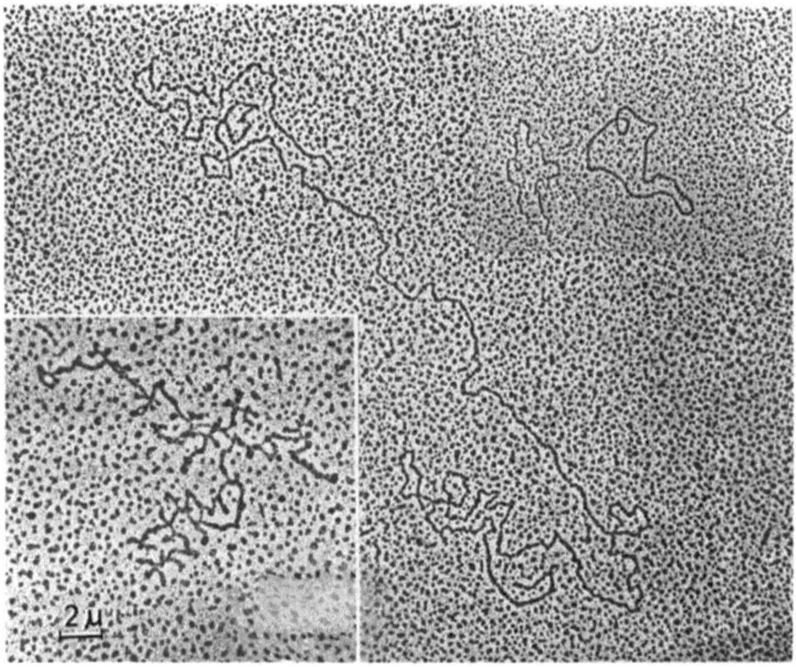

Fig. 3. A single-stranded linear DNA molecule (λ) prepared by the formamide basic protein film technique (40% formamide). Insert at lower left: single-strand linear λ DNA molecule prepared with 10% formamide, which does not remove all of the random base interactions. Insert at upper right: a double-strand circular molecule (ΦX174 RFII) and a single-strand circular molecule (ΦX174) prepared by the formamide basic protein film technique using 40% formamide. Grids were shadowed with Pd/Pt

which is preferably stored in the dark, contains a concentration of uranyl acetate of 5×10^{-5} M in 90% ethanol and is freshly prepared (within 1 h) by diluting a stock aqueous solution of 5×10^{-2} M uranyl acetate in 50 mM HCl. The contrast can be increased by rotary-shadowing with heavy metal but this is not recommended when single strands are present. The grain size for stained DNA is very small and single-strand bushes have the same stain density as duplex DNA. As a result single-strand bushes are easily identified when stained, but are often obscured by shadowing. Excellent contrast can be obtained with the aqueous technique by staining alone.

II. Formamide Technique

For most purposes the spreading solution should contain 0.5 μg/ml of DNA, 0.05 mg/ml cytochrome c, 0.1 M Tris (from Tris-OH+HCl to give pH 8.5), and 10 mM Na_3EDTA in 40% formamide. Formamide solutions should be well buffered since they drift toward a pH of about 5. The initial rate of decomposition of formamide solutions is directly proportional to the hydroxyl ion concentration. Thus, more alkaline buffers such as carbonate cause a more rapid decomposition. The concentration of formamide needed to half denature λ DNA at this salt concentration and in the presence of cytochrome c is about 85%. One can use 50 – 70% formamide when it is desirable to minimize duplex formation due either to relatively short sections of complementary base pairs or to longer sections of DNA that are less homologous because they contain mismatched bases. The hypophase is 30% less in formamide than the spreading solution, 10 mM Tris (pH 8.5), 1 mM Na_3EDTA, and should be made up shortly (5 min or less) before use since it may become acidic relatively quickly. The spreading solution should be used within an hour or two. A clean glass slide, pretreated with 0.25 M NH_4Ac and allowed to drain dry, is put into the petri dish containing the hypophase. Little or no talc need be used since the formamide causes the formation of a large denatured film. Fifty μl of the spreading solution is slowly applied from a glass micropipette at the slide-solution boundary. To help spread out the single-stranded DNA the film is allowed to stand for 30 s to 5 min before being picked up near the slide-solution boundary. Some increase in contrast is obtained if the film is slightly compressed with a Teflon bar before being picked up. In general with the formamide technique, staining alone does not give adequate contrast except for dark field electron microscopy. However, the contrast for light field microscopy is adequate if the DNA is either stained and shadowed, or just shadowed.

III. Reagents

Other authors report that several small globular basic proteins, including cytochrome c, lysozyme, chymotrypin, and trypsin are suitable for forming the protein film that binds the DNA (KLEINSCHMIDT, 1968), but our own experience is mainly with cytochrome c. Dissolve 10 to 20 mg of cytochrome c (Calbiochem, equine heart, salt free, A grade) per ml of distilled water. Allow the solution to stand at 5° C for several days so that most of the nucleic acid contaminants will precipitate. Filter through a Millipore HA, 0.45 μ filter that has been extensively washed with distilled water. The concentration of the stock solution is then measured spectrophotometrically where A_{410} of a 1 mg/ml solution is about 10. This solution is diluted to 1 mg/ml with distilled water and stored at 5° C or salt is added to make solutions containing 2 M Tris-HCl pH 8.5, 0.2 M Na_3EDTA, 1 mg/ml cytochrome c or 5 M NH_4Ac, 1 mg/ml cytochrome c. These solutions are quite stable and can usually be used for several years without loss in performance. We pick up the protein-DNA mixed film on grids covered with a freshly spread (3 – 48 h) parlodion film from a 3.5% w/v solution in n-pentyl acetate. If the parlodion contains any moisture the films will have many small holes. Drying of solid parlodion strips (Mallinckrodt) is accomplished in a vacuum dessicator at 25° C for several days. The pentyl acetate is dried with molecular sieves, and the 3.5% parlodion solution is stored over molecular sieves. Parlodion-covered grids are prepared as follows: A wire screen is placed in a shallow metal container equipped with a drain. The container is filled with distilled water above the level of the screen. Copper grids (200=mesh) (purchased from Pellco) are cleaned free of dirt and grease by dipping into actone and then placed on the surface of the wire screen below the surface of the water. Several drops of the parlodion solution are used to clean the surface of the water; after the pentyl acetate has evaporated the parlodion film is stripped off to remove dirt and grease. One drop of parlodion solution is applied to the surface, and after the pentyl acetate has evaporated, or after about one min, the water is drained out of the container, allowing the parlodion film to settle onto the grids. The screen is blotted with a paper towel to remove the excess water and allowed to dry 2 h before the grids are used. Support films prepared in this manner are quite strong and generally withstand the electron beam at crossover.

IV. Problems Related to Contrast

To measure the length of a DNA molecule, particularly one that has structural features such as deletion or substitution loops, it must be fairly well extended with few crossover points and it must be well contrasted. These features are very dependent on what appear to be minor variations in technique: the amount of talc used, the way the hyperphase is allowed to run down the glass ramp, whether the supporting film is carbon or parlodion, the way the protein film is picked up, the position with respect to the ramp from which the film is picked up, and the length of time the film is allowed to stand, as well as the obviously important features of the composition of spreading solution and hypophase.

The contrast with either staining or shadowing depends on the amount of protein collapsed around the DNA, which itself depends on many factors. The duplex DNA structure observed in the electron microscope varies between 80 and 150 Å in diameter. Two major variables are the ionic strengths of the spreading solution and the hypophase. Less protein collapses around the DNA at lower ionic strength. In the aqueous technique, increasing the ammonium acetate concentration increases contrast, but if the concentration in the hypophase is above 0.5 M, little or no DNA sticks to the film. Furthermore, there is an increasing tendency for lateral aggregation to occur when the hypophase concentration is raised above 0.4 M. Flower-like aggregates sometimes form below 0.2 M NH_4Ac in the hypophase. Therefore, we use 0.5 M and 0.25 M NH_4Ac for spreading solution and hypophase, respectively. Flower-like and other aggregate formations can frequently be decreased by decreasing the cytochrome c concentration. In the formamide technique, there is less tangling of single-stranded DNA at low ionic strengths. Most denaturing agents reduce the contrast; formamide raises it. Good results have been achieved with 85% formamide for double-stranded DNA and up to 95% formamide for single-stranded DNA. However, at high concentrations of salt and formamide the DNA becomes very thickly covered with cytochrome c and lateral aggregation can occur. In our experience, the following reagents and conditions generally cause deleterious effects in the basic protein film technique: (1) acidic conditions [depurination ($pH < 7$) and aggregation ($pH < 6$)]; (2) basic conditions [poor contrast and aggregation ($pH > 10$)]; (3) detergents (aggregation); (4) nitrate or phosphate salts (poor contrast); (5) many denaturing agents (poor contrast); (6) excessive protein in the DNA sample (poor contrast). We have found that recent commercial preparations of ethanol frequently contain some impurity that greatly reduces contrast, and we now routinely redistill our ethanol.

V. Double-Strand/Single-Strand Distinction and Length Ratios

The distinction between double- and single strands is greater when there is less protein collapsed around the DNA, thus at low ionic strength and with just enough formamide to extend the single strands. Shadowing is better than staining for making this distinction.

Increased thickness of single-stranded DNA is correlated with decreased length; thus the single-strand/double-strand length ratio decreases with increasing salt concentration, and with increasing formamide concentration. There may also be a decrease in this length ratio with an increase in the length of the DNA. The single-strand/double-strand length ratio for F factor

Table 1

Denaturing agent		Salt conc.	Single-strand / Double-strand
Formaldehyde		0	1.05
Formamide	5%	0	1.10
Formamide	5%	.01	0.88
Formamide	10%	.01	0.74
Formamide	20%	.01	0.68
Purine	1 M	.1	0.50
CH_3HgOH		.2	0.38

is less than that for adenovirus DNA, which in turn is less than that for ΦX174 DNA. The length ratio may also be affected by base sequence but this has not been tested. Single strands are very susceptible to stretching and therefore the single-strand/double-strand length ratio will increase if the single strands are stretched. Table 1 gives single-strand/double-strand length ratios under several different conditions. The values given are for one set of experiments. Somewhat different values are obtained for different preparations even under apparently identical conditions.

C. Heteroduplex Molecules

A type of DNA molecule that has come to be of great usefulness in electron microscopy for a variety of mapping purposes is the heteroduplex molecule. As its name implies, a heteroduplex molecule is constructed by annealing together two related yet different single strands of DNA to form a duplex structure. Where the two single strands contain complementary bases the heteroduplex molecule will be double-stranded, and where the two single strands contain different bases, the heteroduplex molecule will consist of 2 unpaired single strands. By visualizing the molecule in the electron microscope under conditions that allow discrimination between double- and single-stranded DNA, the regions of base complementarity and noncomplementarity can be identified. Under suitable conditions these regions can be measured and mapped. Experimentally, a heteroduplex molecule is constructed by mixing together the two DNA samples of interest, denaturing them, and then allowing them to renature. The reaction mixture will consist not only of the heteroduplex molecules, but also of renatured parent homoduplex molecules and unrenatured single-stranded molecules. However, both of these are easily distinguishable from the heteroduplex molecules and their presence in no way interferes with the experiment.

I. Experimental Procedure

Renaturation in aqueous solution requires heating to about T_m –30° C, (T_m=melting temperature) a precedure that causes single-strand breaks. A short heating procedure (20 s) at a high DNA concentration ($A_{260}\approx 1.0$) and in the presence of high salt has been successfully used (DAVIS and DAVIDSON, 1968). Renaturation in formamide is a better procedure (WESTMORELAND et al., 1969). One may use moderately dilute DNA solutions ($A_{260}\approx 0.1$), room temperature or less, and longer times. Some data on the effects of temperature and ionic medium on the thermodynamics and kinetics of renaturation in formamide are available (MCCONAUGHY et al., 1969;

MASSIE and ZIMM, 1969). The effects of complexity, molecular weight, temperature, and ionic strength can also be estimated from studies of renaturation kinetics in aqueous solution (WETMUR and DAVIDSON, 1968; STUDIER, 1969).

The detailed description below applies to lambdoid phage DNAs. Denatured DNA is prepared directly from the phage by simultaneous lysis and denaturation in alkali. The presence of the phage protein, or phage ghosts, which may occasionally be seen on the grid, does not interfere with electron microscopy. For example, 5×10^9 particles (0.25 μg DNA) of each of the two phages are placed in a 6×50 mm glass tube. Distilled water is added to a final volume of 39 μl. One μl of 1 M Na_4EDTA (made by adding 2 mol of NaOH to 1 mol Na_2EDTA/l of solution) is added followed by addition of 5 μl 1 M NaOH. This causes release of the DNA from the phage particle and denaturation of the DNA. After 10 min at room temperature, the DNA solution is neutralized by the addition of 5 μl of a solution containing 1.8 M Tris-HCl [Tris(hydroxymethyl)aminomethane hydrochloride] and 0.2 M Tris base [Tris(hydroxymethyl)aminomethane]. The pH usually reads 8.5. Renaturation is accomplished at room temperature (25° C) by the addition of 50 μl formamide. These conditions are about T_m −30° C, which gives an optimal rate. At lower temperatures renaturation is much slower. At high

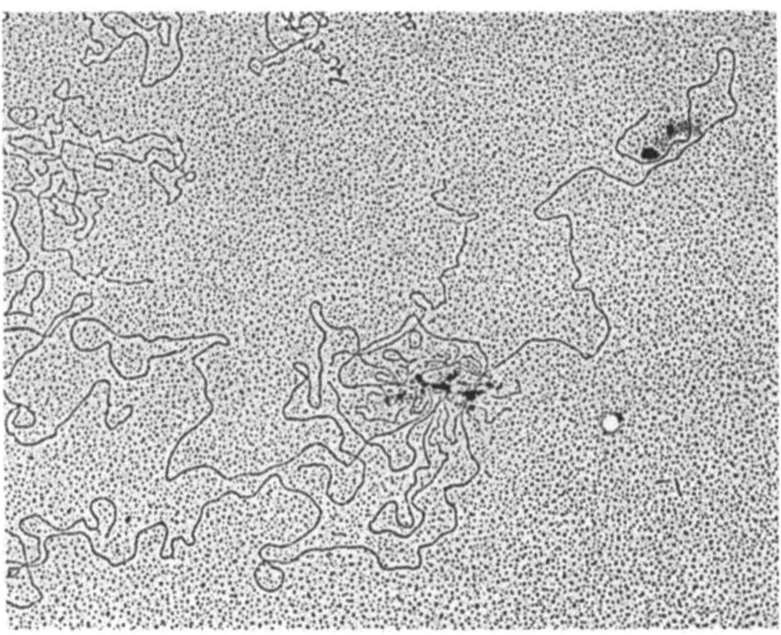

Fig. 4. A renatured tangle of several λ DNA molecules

temperatures (35° C) the rate is about the same, and the more stringent renaturation conditions appear to reduce the number of tangled renatured molecules (Fig. 4). About 50% renaturation occurs in ½ to 1 h, and the pH is then usually between 7.5 and 8.0. The formamide solution is cooled to 0° C to slow down the renaturation reaction, and immediately mounted for electron microscopy.

II. Examples

Application of the heteroduplex technique has found wide-ranging uses. Figure 5 shows an example of the heteroduplex technique as it has been used to map the position and length of deletion and substitution mutations in λ DNA. A micrograph such as that of the 434/λ heteroduplex is interpreted to mean that in the regions that are double-stranded, 434 DNA and λ DNA have the 'same' base sequence and in the regions where single-stranded loops occur, the base sequences are 'nonhomologous'. It must be emphasized that

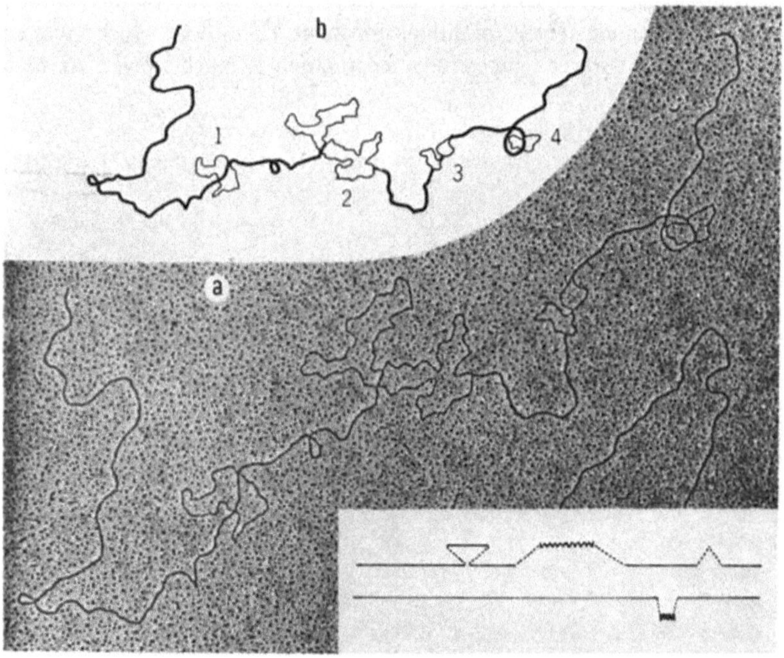

Fig. 5. (a) A heretoduplex molecule between λa200b189*bio*69*c*I857*nin*5 DNA and λ*imm*434 DNA (THOMAS et al., 1974) mounted in 40% formamide. (b) Illustration of micrograph in (a). *Region 1* a200 duplication; *region 2* b189*bio*69/λ nonhomology; *region 3 imm*434/λ nonhomology; *region 4 nin*5 deletion. *Insert in lower right* schematic illustration of heteroduplex molecule

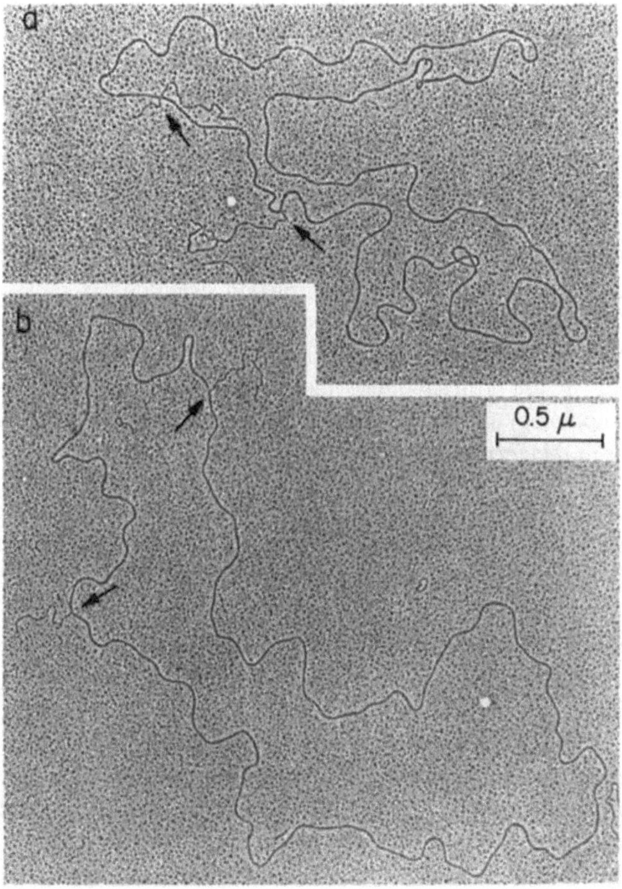

Fig. 6 a and b. Circularly permuted, terminally repetitious coliphage 15 DNA molecules that have been denatured and renatured. Mounted in 40% formamide and shadowed with uranyl acetate

at present we do not know how perfectly complementary two sequences must be to give a double-stranded structure, and how little complementarity there must be before one sees loops and concludes that the base sequences are nonhomologous.

Heteroduplex techniques can be used to demonstrate that certain bacteriophages have linear DNA molecules that are circularly permuted and terminally repetitious. When such a DNA molecule is denatured and renatured, circular duplex structures with protruding single-stranded branches are formed (Fig. 6; THOMAS and MACHATTIE, 1964). If the renatured molecules are spread in the presence of formamide, the length of the terminal repetition and its length heterogeneity can both be measured (LEE et al., 1970). Although thus far only examples of bacteriophage DNA heteroduplexes have been mentioned, heteroduplex techniques have also proved useful in studying deletion, substitution, and duplication mutations in animal viral DNAs such as that of simian virus 40 (SV40) (TAI et al., 1972; RISSER and MULDER, 1974; MERTZ et al., 1974). SV40 mutations containing a tandemly duplicated region around the origin of DNA replication have been isolated and characterized by heteroduplex mapping (BROCKMAN et al., 1974). Heteroduplex techniques have also been used to map the sites of initiation of transcription by RNA polymerase in vitro, by constructing heteroduplex molecules between in vitro synthesized RNA and the DNA template (DAVIS and HYMAN, 1970). A special class of heteroduplex molecules called diheteroduplexes, and the use of heteroduplex molecules to study in depth the partial sequence homology of related DNA molecules, will be discussed in more detail later in this Section.

III. Complications Which May Arise in Constructing and Examining Heteroduplex Molecules

When heteroduplex molecules are examined in the electron microscope, a major cause of uncertainty in interpretation is the presence of fragments due to single-strand breaks. When two intact, closely related single strands renature, the resulting molecule is largely double-stranded and does not renature further. When fragmented strands renature, the product is more likely to have a sufficient length of exposed single strands so that further renaturation to give branched and generally misleading aggregates occurs (Fig. 7). Thus length measurements and quantitative mapping are very difficult if one does not have intact strands of DNA. The procedures for DNA preparation, denaturation, and renaturation should therefore be designed to minimize the number of single-strand breaks (DAVIS et al., 1971). To further minimize the problems due to higher aggregates even with intact DNA strands, we recommend under-renaturing the DNA. Renaturation of approximately 50% is optimal. Unrenatured single strands of DNA are easily recognized and can be ignored. For most purposes there is no advantage in separating complementary strands (WESTMORELAND et al., 1969) and we recommend avoiding the additional manipulations involved.

For heteroduplex experiments starting with closed circular DNA molecules, it is desirable to introduce either one single-strand break or one double-strand break per molecule. The former can be accomplished by lightly nicking the separated closed circular molecules with DNase (HUDSON et al., 1969) or with DNase in the presence of ethidium bromide (GREENFIELD et al., 1975). The latter is most easily accomplished with the use of restriction

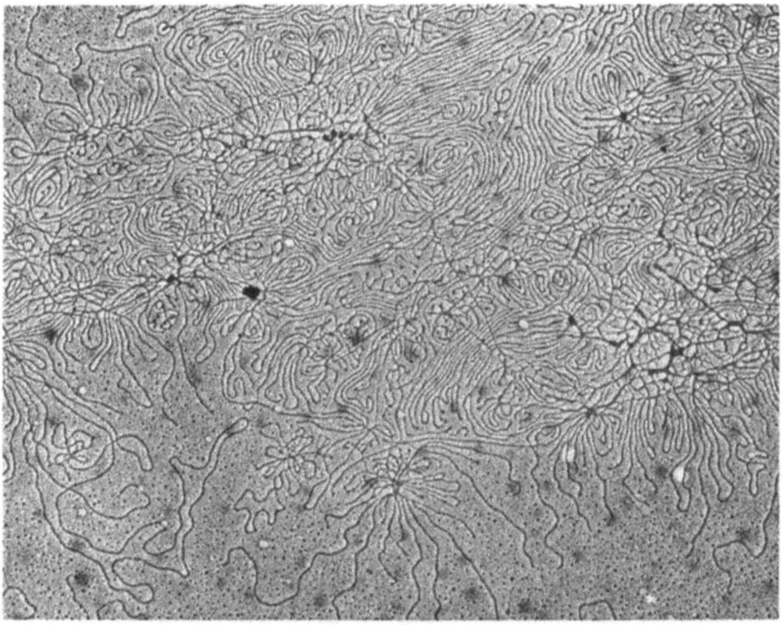

Fig. 7. A renatured aggregate of DNA resulting from over-renaturation

endonucleases. The use of restriction enzymes has greatly increased the ease with which it is possible to do all kinds of electron-microscopic mapping of DNA molecules. It is possible to cleave DNA molecules at specific sequences (KELLY and SMITH, 1970; HEDGPETH et al., 1972; BIGGER et al., 1973) and to rejoin cleaved molecules with *E. coli* DNA ligase (MERTZ and DAVIS, 1972) by virtue of overlapping single-strand ends. For instance, SV40 DNA is cleaved once with the restriction endonuclease *Eco*RI (MORROW and BERG, 1972); thus it is possible to prepare linear SV40 DNA with defined end points for the purposes of constructing heteroduplexes by a cleavage step with *Eco*RI.

The excluded volume effect influences various heteroduplex experiments, and its implications in any given experiment should be considered (DAVIS et al., 1971).

IV. Branch Migration

In a heteroduplex molecule that contains a single-strand tail having the same base sequence as an adjacent strand in the duplex, the single-strand tail can displace this adjacent strand in the duplex by 'branch migrating' (LEE et al., 1970). As shown in Figure 8, the displaced strand is now seen as a single-strand tail. Intermediate stages of displacement, where the heteroduplex

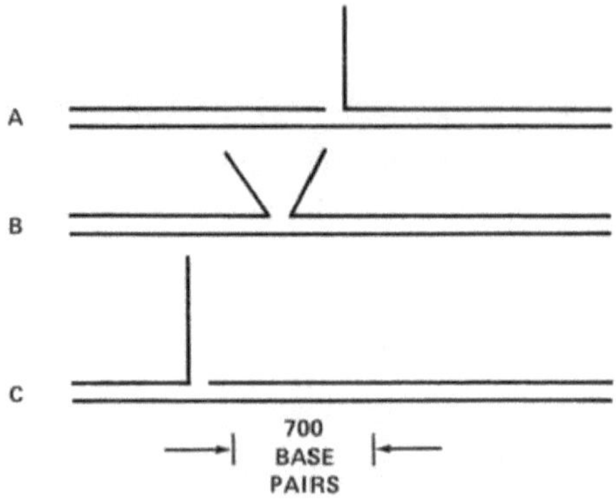

Fig. 8. A – C. Schematic representation of branch migration. See text for explanation

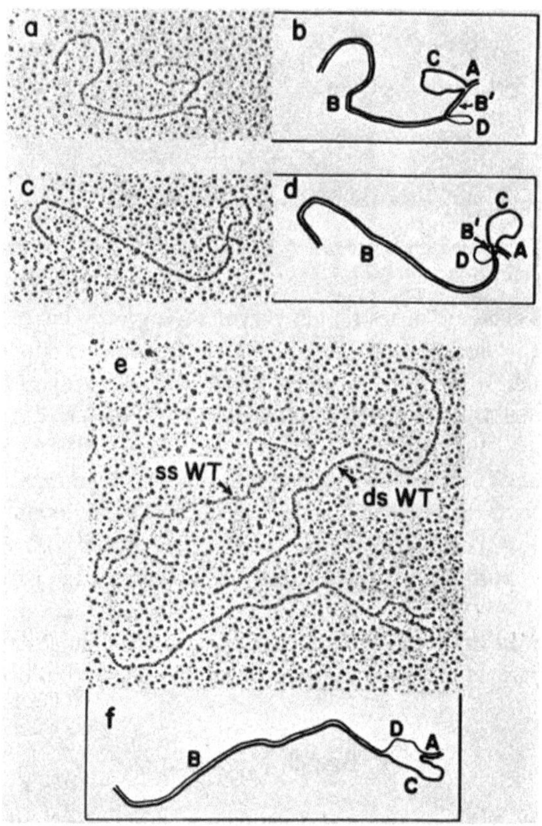

Fig. 9. a – f. SV40 mutant dl-833a heteroduplexed to its parent strain WT830. b, d, and f are illustrations of the heteroduplex molecules seen in a, c, and e, respectively. Segment *B'* is a tandem duplication of segment *D*, thus loop *D* can branch migrate along the length of segment *B'*. In e, the *C* loop and the *D* loop have merged. This figure has kindly been provided by Dr. JANET MERTZ

molecule has two single-strand tails, are also observed. This is the phenomenon of single-strand branch migration, which has been postulated to perform a role in in vivo recombination (CASSUTO and RADDING, 1971). Double-strand branch migration has also been observed in vitro (KIM et al., 1972) and postulated to occur during in vivo recombination (BROKER and LEHMAN, 1971). At all stages of branch migration the molecule contains an equivalent number of hydrogen-bonded base pairs; thus, branch migration is not an unexpected phenomenon based on physicochemical considerations. An example of single-strand branch migration is seen in Figure 9.

V. Terminology, Topology, and Stability of Branch Points

Because of the large variety of branched molecules that have been observed in the electron microscope, we feel that it is useful to introduce a terminology for branches in DNA. Figure 10 contains a list of commonly observed

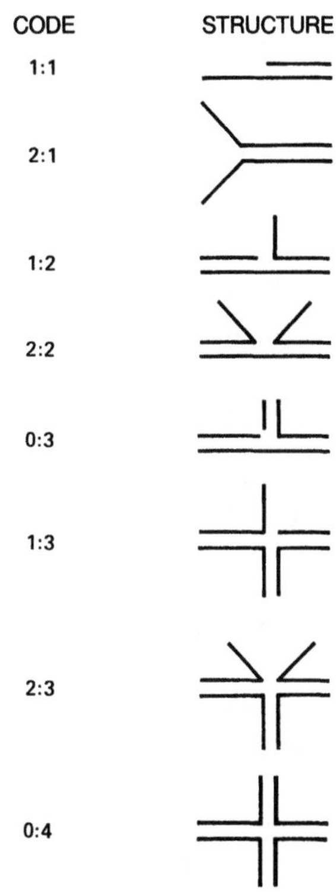

Fig. 10. Commonly observed branch structures in DNA. The code used to identify branches is the number of single strands in the branch followed by the number of double strands

branches and a suggested code. This code represents simply the number of single strands in the branch followed by the number of double strands. Thus, the branch formed by a deletion loop in a heteroduplex molecule would be a 2 : 2 branch, while the branch formed by a substitution loop would be a 2 : 1 branch. An important feature of branch point topology is the fact that a duplex DNA strand may or may not have a swivel at the branch point such that it can rotate if the other strands are fixed (Fig. 11). All single strands are free to rotate.

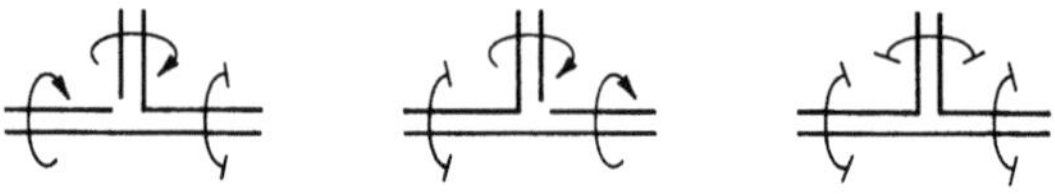

Fig. 11. Swivels at branch points in DNA. The first two branch structures contain swivels that allow free rotation of two of the three stands, while the third branch structure is constrained from rotation in all three strands

The branches shown in Figure 10 are not all equally stable. If, through branch migration, different branch forms are interconvertible, then knowledge of the relative stabilities allows one to predict which form will predominate. For example, in the simple branch migration model shown in Figure 8, the three forms A, B, and C are found with equal frequency (LEE et al., 1970). The 2 : 2 branch (from B) has a statistical weighting factor of 700 because the branch can occur at any one of the 700 base pairs. The 1 : 2 branch (forms A and C) has a statistical weighting factor of only 2. Since all three forms are found in equal numbers, the 1 : 2 branch must have a stability factor of 350 greater than that of the 2 : 2 branch. This corresponds to a difference in free energy of about 3500 cal/mol in favor of the 1 : 2 branch, which presumably results from differences in the steric interactions at the junction of the DNA strands in 1 : 2 and 2 : 2 branches.

There are more complicated interconvertible branch structures involving the probability of ring closure that are elegantly discussed by KIM et al. (1972).

VI. Diheteroduplexes

The construction of diheteroduplexes (sometimes referred to as triple heteroduplexes or heterotriplexes) can occasionally provide useful information. As the name implies, these structures are formed by annealing three different single-stranded DNAs or combinations of DNA and RNA together to form two (*di-*)heteroduplex structures in the same molecule. They are limited in usefulness by their low frequency of occurrence, making the measurement of many molecules unfeasible. However, they are often used to determine directionality or to confirm a proposed structure, cases where a small number of molecules is sufficient to answer a specific question, rather than for doing detailed mapping.

The construction of diheteroduplexes made it possible to orient the location of the SV40 DNA sequence carried in the adeno-SV40 hybrid DNA, $Ad2^{+}ND_1$, relative to the *Eco*RI restriction endonuclease cleavage site in SV40 DNA, and to determine which of the two single-stranded segments in the $Ad2^{+}ND_1$-Ad2 heteroduplexes is the SV40 DNA segment (MORROW et al., 1973). Figure 12 shows an example of a diheteroduplex molecule be-

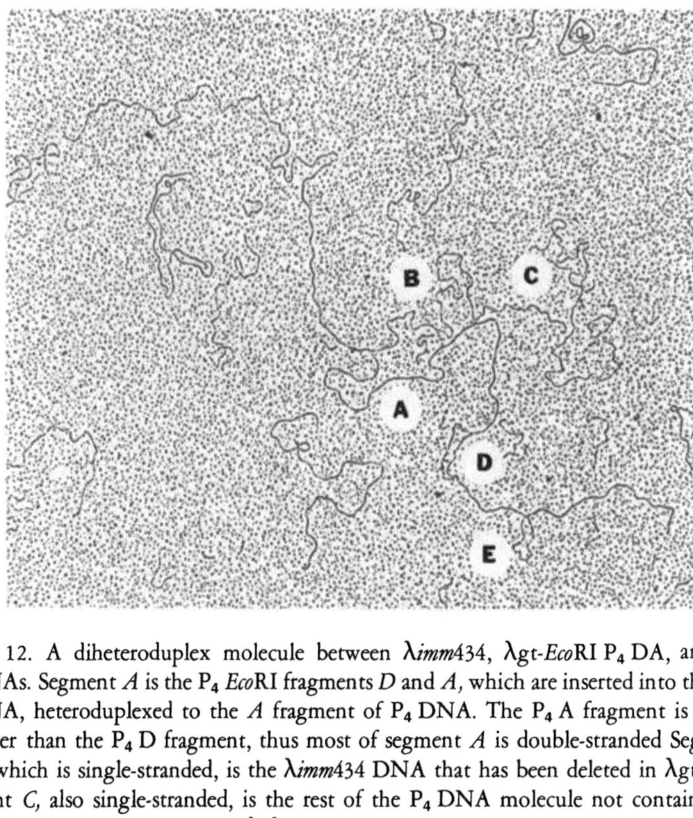

Fig. 12. A diheteroduplex molecule between λ*imm*434, λgt-*Eco*RI P_4 DA, and P_4-DNAs. Segment *A* is the P_4 *Eco*RI fragments *D* and *A,* which are inserted into the λgt DNA, heteroduplexed to the *A* fragment of P_4 DNA. The P_4 A fragment is much larger than the P_4 D fragment, thus most of segment *A* is double-stranded Segment *B,* which is single-stranded, is the λ*imm*434 DNA that has been deleted in λgt. Segment *C,* also single-stranded, is the rest of the P_4 DNA molecule not contained in segment *A.* Segment *D* is the λ/λ*imm*434 immunity nonhomology region. Segment *E* is the *nin*5 deletion loop contained in λgt DNA. This micrograph was kindly provided by ELIZABETH RALEIGH

tween λ *imm* 434, λ gt-*Eco*RIP$_4$DA, and P$_4$DNAs. The construction of this dihetero duplex provided unequivocal proof of the structure of the λ transducing phage carrying the *Eco*RID and A fragments of bacateriophage P$_4$DNA (RALEIGH and DAVIS, in preparation).

VII. Partial Sequence Homology

A detailed study has been made of the DNA base sequence homology between coliphages T7 and T3 (DAVIS and HYMAN, 1971). The regions of homology and nonhomology were mapped on the genome, and an estimate of the *degree* of sequence homology in the homologous regions was made. Degree of sequence homology, or percent sequence homology, refers to the fraction of bases in a given segment that are mismatched and thus cannot

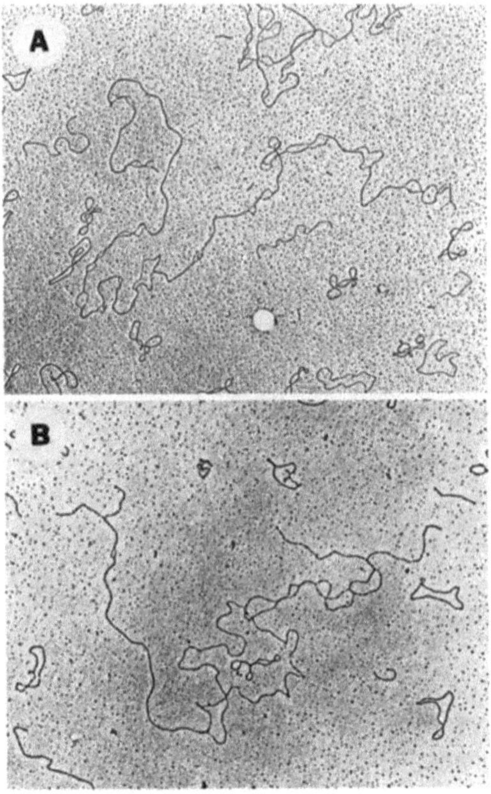

Fig. 13. A and B. T3/T7 heteroduplex molecules. The molecule in (A) is at T_m −28° C. The sample was mounted from a solution containing 45% formamide (cation concentration 0.06 M) onto a hypophase containing 15% formamide (cation concentration 0.006 M). The small circular molecules are ΦX174 DNA. The molecule in (B) is at T_m −6° C. The sample was mounted from a solution containing 75% formamide (cation concentration 0.06 M) onto a hypophase containing 45% formamide (cation concentration 0.006 M). Note the large regions of nonhomology. Note also that the DNA is much thicker in appearance when mounted under higher formamide concentrations

base pair. This is a rigorous definition and should not be confused with other uses that appear in the literature. 'Degree of homology' is often used to refer to the fraction of the total length of a DNA molecule that is measured as duplex under a given set of experimental conditions. This, however, is not an intrinsic property of the DNA molecule. The true degree of sequence homology is reflected in the melting temperature of the DNA segment, as unpaired bases destabilize the helix structure. Thus to estimate the degree of sequence homology in a given DNA segment one can study the melting characteristics of that region. This requires that one be able to control carefully the effective temperature (i. e., the temperature relative to the melting temperature of the DNA under a given set of conditions) of the DNA molecule during the preparation procedures. If the effective temperature were to change during the spreading procedure, the molecule would no longer be at thermal equilibrium and an interpretation of the degree of melting of any given region would be meaningless. One can achieve any given effective temperature, say T_m −30°, by a variety of different conditions. Formamide lowers the melting temperature (T_m) of DNA (MCCONAUGHY et al., 1969) and is used to obtain denaturing conditions. The exact concentration of salt and formamide, as well as the temperature, must be known, for all of these affect the melting temperature of DNA. A procedure, called isodenaturing, has been developed for controlled effective temperature spreading (DAVIS and HYMAN, 1971). In this procedure the denaturing conditions of the spreading solution are identical to those of the hypophase. Since, of necessity, the salt concentrations are different between spreading solution and hypophase, the formamide concentrations must be appropriately different to compensate for the effect of salt concentration on the melting temperature of DNA. Figure 13 shows micrographs of two heteroduplex molecules between T3 and T7 DNA prepared at two effective temperatures. Using an empirical relationship between formamide concentration and the melting temperature of DNA (MCCONAUGHY et al., 1969), it is possible by measuring many heteroduplex molecules prepared at different effective temperatures to construct a percent sequence homology map of the T3 and T7 genomes. A comparison of this map with the genetic maps of T3 and T7 reveals important insights into the evolution of these two phages.

Another variation on this same theme is the study of the weak sequence homology between the DNAs of the two animal viruses, SV40 and polyoma (FERGUSON and DAVIS, 1975). In this case such a small amount of sequence homology exists that it is not detectable by conventional hybridization methods (CUZIN et al., 1970). However, a 15% region of weak sequence homology is revealed by an electron-microscopic method. The procedure involves the construction of hybrid DNA molecules containing an SV40 DNA molecule covalently joined to a polyoma DNA molecule, using the sequential action of *Eco*RI restriction endonuclease and *E. coli* DNA ligase. Denaturation of such hybrid molecules then makes it possible to examine *intra*molecularly rather than *inter*molecularly renatured molecules. Four examples of such intramolecularly renatured 'snapback' molecules with duplex regions of homology are seen in Figure 14.

This study demonstrates the way in which it is possible to take a rigorous, quantitative approach to an electron-microscopic problem. It is likely

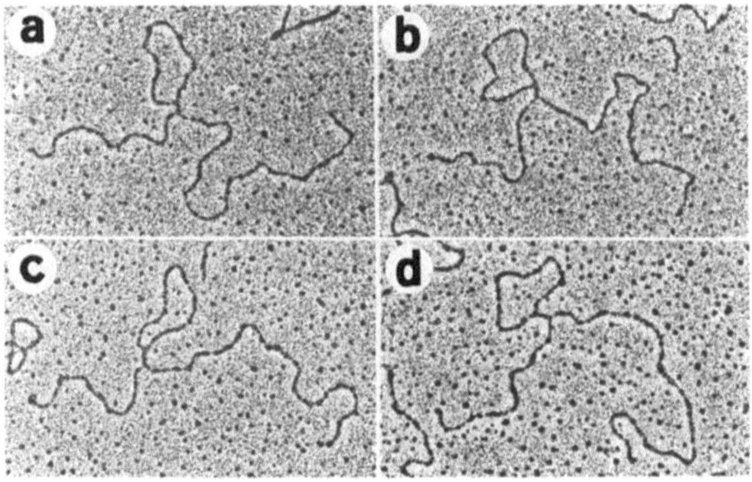

Fig. 14. a – d. Four examples of intramolecularly renatured SV40-polyoma hybrid DNA molecules prepared at T_m –35° C. In each case a short duplex region is observed

that molecules with a very small duplex region of homology (200 base pairs or less) would be overlooked as single-stranded molecules containing an overlap. Therefore, to determine exactly the location of all possible homology regions and to eliminate any subjective selection of snapback molecules, the data were analyzed by systematically scanning the grid and taking pictures of all molecules containing any kind of overlap or duplex region. The position of the overlap was measured in each molecule, and plotted in such a way as to display the position and length of each overlap. A very small region of homology is distinguishable from a simple overlap by the fact that a homology region must map at a unique location. A control experiment was done to determine the distribution of random overlaps in a population of single-stranded DNA molecules that did not have any homology regions. The distribution that was observed results from a small loop containing approximately 1500 bases (15% of the length of the molecule) that can be located with equal frequency anywhere along the length of the molecule. This may be a result of the manner in which an overlapped molecule is stretched while being constrained to the two-dimensional spreading protein film during preparation for microscopy, and is referred to as the 'garden hose' effect.

VIII. Partial Denaturation Mapping

Partial denaturation mapping using alkali and formaldehyde fixation has been elegantly developed and used by INMAN and SCHNOS (INMAN, 1967; INMAN and SCHNOS, 1970; SCHNOS and INMAN, 1970). WOLFSON et al., (1971) developed a simpler method using high concentrations (about 80%) of formamide that gives less reproducible denaturation patterns. A difficulty encountered in using such high formamide concentrations is lateral aggregation of denaturation bubbles (Sect. E II). This problem can be reduced by lowering the salt concentration in the spreading solution to 50 mM Tris (pH 8.5) and 5 mM Na_3EDTA. The main difficulty with lowering the salt concentration of the hypophase is that there is salt in the formamide (probably ammonium formate). Ninety-nine percent (99%) MCB formamide usually contains about 10 mM salt. The hypophase contains about 50% formamide (30% lower formamide concentration than the spreading solution), which thus contributes about 5 mM salt, which is the desired salt concentration. Therefore, the hypophase contains just formamide in distilled water. The pH of the unbuffered hypophase is about 5 and actually improves the method by reducing the thickness of the DNA strands. Since this reaction is at equilibrium, unlike the formaldehyde method, the exact degree of denaturation desired can be achieved by continuing to adjust the formamide concentration of the spreading solution. The volume of the spreading solution used is sufficient for a number of trial preparations. The uranyl acetate staining method gives adequate contrast and each grid can be checked immediately after it is prepared.

D. Measuring and Error Analysis

I. Measurement Procedures

Our measurements are made with a Hewlett-Packard 9864A Digitizer and 9810A Calculator with a fully smoothed length calculation program giving an accuracy of 0.5% and a greater degree of reproducibility on sample figures of known length. The 35-mm negative (Kodalith estar base film LR Film 2572) is projected with a Prado Universal projector up, across, and down with the aid of two surface-silvered mirrors onto the digitizer surface. The total path length of the projected image is about 6 feet. The projector is equipped with 120-, 65-, and 35-mm lenses that project with a magnification of about 15, 30, and 60 times, respectively. All measurements are made relative to a standard DNA molecule mounted on the same grid. Lengths are never measured in absolute length units.

II. Reference Markers and Orientation

The use of a physical marker such as a substitution or deletion loop as a reference point on a linear molecule to be measured is often useful for providing internal calibration for mapping a second feature on the molecule or for orienting the molecule with respect to its ends to determine the unequivocal map position of a second feature on the molecule. Restriction endonuclease cleavage sites have proved to be very useful as unique markers on DNA. Mapping deletions or substitutions relative to restriction sites is achieved by cleaving heteroduplex molecules with restriction endonucleases (THOMAS and DAVIS, 1974).

The problem of orientation can be a significant one. An attempt was made to map the preferred binding site of T antigen on SV40 DNA (REED et al., 1975) by comparing histograms of ferritin-binding positions from two separate populations of defined linear molecules, one obtained by cutting circular SV40 DNA with *Eco*RI restriction endonuclease, and the other by cutting with *Hpa*II restriction endonuclease. It was not possible to determine unequivocally the preferred binding site by matching these two separate histogramms because of the high background of ferritins seen on the DNA. The use of a physical marker on the DNA, in this case the small denatured region seen after binding *E. coli* DNA unwinding protein to SV40 supercoiled DNA molecules, was required. With the presence of a physical marker on each molecule, the position of each bound ferritin could be determined unequivocally with respect to the ends of the DNA molecule, and a histogram with a high background but single clear peak at the unique preferred binding position was obtained.

III. Error Analysis

The observed length of the double-stranded DNA molecules in a homogeneous population of molecules all of which contain the same number of base pairs in the same sequence fluctuates about a quantitatively significant mean with a quantitatively reproducible standard deviation. The variance (σ_D^2) around the mean for this intrinsic fluctuation in length is directly proportional to the mean length $\langle L \rangle$ of the DNA. Thus,

$$\sigma_D = D\langle L \rangle^{1/2} \tag{1}$$

The interpretation of this empirical fact is that there is some fundamental segment length for DNA molecules mounted in basic protein films and observed in the electron microscope. The contour length of a molecule is the sum of the segment lengths. There is an intrinsic variability in the length of a segment, and each segment length is a stochastically independent variable. A possible source of this variable is that a segment may take the A or B form as defined by X-ray diffraction of DNA fibers and is a result of the dehydration required for electron microscopy. Eq. (1) asserts that if a set of measurements is made on a sample (*S1*) of n molecules to give a mean length $\langle L \rangle_{S1}$, the standard deviation of the sample (σ_{S1}) should agree with the theoretically derived σ_D. The constant D of Eq. (1) has been experimentally evaluated from the measurement of about 10^4 molecules ranging in length from 200 to 50,000 base pairs, and it was found that D=2.2 when the length is expressed in base pairs. The principles of statistics assert that if sets of measurements are made on a sample, each set with n molecules, then a set of mean lengths will be generated, $\langle L \rangle_{S1}$, $\langle L \rangle_{S2}$, etc. These mean lengths will be normally distributed about the mean of the means $\langle L \rangle m$ with a standard deviation σm. The standard deviation of the means (σm) is the standard error (ϵ) and may also be derived from the standard deviation of a single sample (σ_s) if n is large according to the relationship $\epsilon = \sigma_{s/n}{}^{1/2}$ or from Eq. (1) according to the relationship $\epsilon = \sigma_{D/n}{}^{1/2}$. It is standard error that assesses the certainty of a particular mean and not the standard deviation of the sample (σ_s). These predicted relationships concerning the standard error have been confirmed by experimentation.

The most reproducible length determination of a homogeneous population of DNA molecules is made relative to the length of a calibrating DNA molecule. Therefore, the standard error of this determination ϵx is dependent upon the number of measurements nc and length $\langle Lc \rangle$ of the calibrating DNA molecule, as well as the number of measurements nx and length $\langle Lx \rangle$ of the unknown molecule and is given by the relationship

$$\frac{\epsilon x}{\langle Lx \rangle} = D \left(\frac{1}{nx \langle Lx \rangle} + \frac{1}{nc \langle Lc \rangle} \right)^{1/2} \tag{2}$$

IV. Determination of Number Average Molecular Weight

The number average molecular weight of a DNA sample is readily determined by electron microscopy. It has always been assumed that the populations of DNA molecules observed in an electron microscope field accurately reflect the populations in the solution, or more specifically, that the size of a DNA molecule does not affect its probability of being found on the grid. This assumption has recently been confirmed (RALEIGH and DAVIS, 1976).

There are three things one must consider in making a number average molecular weight determination:

1. DNA Standard

An internal DNA length standard must be included. This standard molecule must be readily distinguished from the population of unknown molecules, and its molecular weight must be accurately known. Circular molecules are quite useful as standards for linear molecules.

2. Unbiased Sampling of Molecules

In selecting molecules to be measured in a heterogeneous population, there may be bias against long molecules that are more likely to extend out of a given field than small ones. This bias is largely avoided by the end scoring procedure. This is accomplished by projecting a 35-mm negative onto the digitizing surface and marking a scoring frame with a defined field smaller than the projected negative. The scoring frame must be marked such that all molecules extending out of this frame can be measured. All molecules totally within the scoring frame are measured. Any molecule extending out of the scoring frame is measured if both ends lie within the frame. If neither end lies within the frame, it is not measured. If one end lies within the frame, it is measured but given a weighting factor of 1/2 (scored as 1/2 molecule and 1/2 the length in determining the mean).

3. Background Subtraction of Contaminating DNA Molecules

Almost all reagents used for electron microscopy (cytochrome c, standard DNA, etc.) are contaminated with small fragments of DNA, or material that appears as DNA in the electron microscope. These DNA fragments are generally a few hundred base pairs long.

Table 2. DNA length ratios

T7/λ	0.824
T3/λ	0.825
P4/λ	0.236
ΦX174 R.F./λ	0.110
PM2/λ	0.215
SV40/λ	0.1065
434/λ	1.00
21/λ	0.890
80/λ	0.920
82/λ	0.915

This correction need not be considered if large molecules are being measured. The size and number of these background fragments can be determined from a grid prepared with the standard DNA alone. The number of contaminating fragments to be subtracted from the unknown population can be determined by normalizing relative to the numbers of standard DNA molecules.

Table 2 includes some commonly used DNA length ratios.

V. Determination of DNA Concentration by Electron Microscopy

It is possible to determine the concentration of DNA in a sample of defined molecular weight by mounting it in the presence of a distinguishable standard of known molecular weight and concentration (RALEIGH and DAVIS, 1976). As little as 5 ng of DNA are needed for this determination, and the concentration of a given species can be determined directly in the presence of a large excess of extraneous DNA if the given DNA is distinguishable (e.g., a small circular molecule in the presence of a large amount of linear DNA).

Counts of DNA populations are made by projecting 35-mm negatives onto a scoring frame smaller than the projected negative such that all molecules extending out of the field can be classified by size and shape. Ends are counted; thus, a molecule falling partially within the scoring frame is counted as a whole molecule, a half, or not at all, depending on whether both, one, or neither end falls within the scoring frame. This scoring method avoids bias against long molecules, which are more likely to extend out of a given field than small molecules. The smaller the difference in size between the standard and unknown DNA molecules, the less essential it is to avoid bias against the longer molecule.

E. Artifacts and Topology Considerations

This Section deals with the interpretation of molecular structures that are unique to the basic protein film spreading method and that probably do not reflect intrinsic properties of DNA molecules in solution. Also, we have included topologic restrictions to renaturation that probably exist in solution but have not yet been described.

I. Flowers

'Flower' is a descriptive term used to refer to an artifactual structure (Fig. 15) that was found in much of the early DNA microscopy. What

Fig. 15. A 'flower' structure that is sometimes observed when viewing DNA in the electron microscope. Flower structures appear to have dense centers around which the DNA molecule aggregates

causes its formation is not clear, but it seems to result from some aggregate of cytochrome c that binds to DNA. Its appearance is totally absent when the cytochrome c solution is prepared as described in Section B III, and the DNA is mounted at the salt and cytochrome c concentrations suggested.

II. Lateral Aggregation

The success of the basic protein film technique results from the fact that no attraction exists between or within DNA molecules to cause aggregation. However, if the DNA becomes thickly covered with protein, the DNA strands can stick together along their length (lateral aggregation). Lateral aggregation occurs at high formamide concentrations, high salt concentrations, and under other spreading conditions that cause the DNA to become very thickly coated with protein. One must be very careful in the interpretation of secondary structures in single-stranded DNA. What is interpreted to be 'snapback' or inverted repeat DNA may only be lateral aggregation. The molecule in Figure 14 shows secondary structure, and not lateral aggregation, because it occurs at a unique location and it melts at higher formamide concentrations. However, such a positive identification cannot be made with a heterogeneous DNA population.

III. Intermolecular Overlap

As discussed in Section E II, there appear to be no attractive forces between DNA molecules when mounted with the basic protein film method. However, there do appear to be mild repulsive forces between molecules. As shown in Figure 16, when the DNA is mounted at very high concentrations, few molecules overlap. Some of this behavior may be due to alignment of molecules with the flow fields. Whatever the reason, this observation means that if two molecules do overlap at low DNA concentrations, then they may be attached to one another (by covalent bonds, hydrogen bonds, cross links, or topologic links).

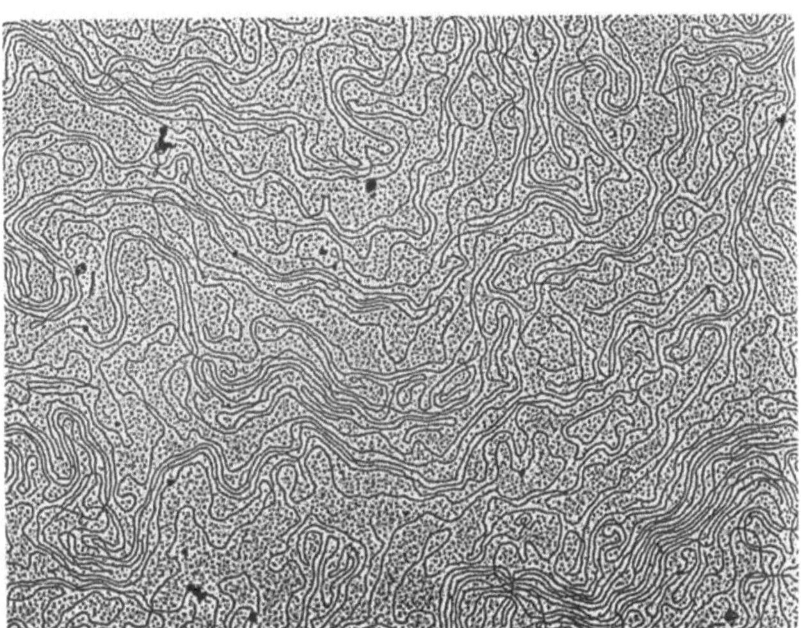

Fig. 16. DNA mounted at high concentration

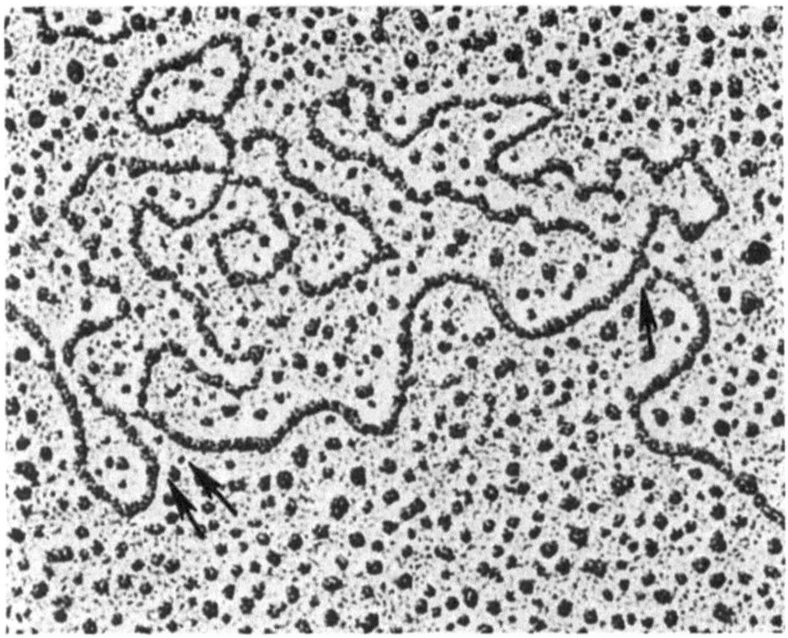

Fig. 17. A heteroduplex molecule between a biochemically constructed λ phage DNA carrying a *Drosophila* satellite DNA fragment of less than 100 base pairs and a tandem duplication of λ DNA (λgt-Dn106), and a genetically selected λ p *ara* phage DNA carrying part of the *E. coli* arabinose operon. The single arrow points to the tandem duplication loop, and the double arrow points to the substitution loop consisting of the *E. coli* arabinose DNA in one strand and the small *Drosophila* insert in the other strand. This micrograph has kindly been provided by Dr. PETER PHILIPPSEN

IV. Branch Peelback in Heteroduplex Molecules

It has frequently been observed that on some, but clearly not all grids, there is local denaturation (peelback) of about 50 to 100 base pairs in most 2 : 2 branches in DNA (see Fig. 10). When such structures are encountered in heteroduplex molecules, it is difficult to decide whether this is a 2 : 2 branch with peelback, or two 2 : 1 branches with a small intervening insertion. Figure 17 shows a heteroduplex molecule with a 2 : 2 branch (tandem duplication), and two 2 : 1 branches separated by a segment of DNA of less than 100 bases. This latter branch structure is very characteristic of peelback at 2 : 2 branches. However, in this case the tandem duplication serves as an internal control branch structure and always appeared on this grid as a 2 : 2 branch. Therefore it is possible to decide in favor of two 2 : 1 branches separated by a small insertion for the other structure. This assignment has been confirmed by other methods. Branch peelback seems to occur on grids that are picked up from films that were not restricted with talc and from films that were allowed to remain on the hypophase for longer than 1 min. These grids are also characterized by having single-stranded DNA that is well extended and with few kinks.

V. 2 : 2 Branch Point Configuration

It has been observed (MERTZ, 1975) that the two single strands from 2 : 2 branches seldom overlap (Fig. 18). One can use this observation for

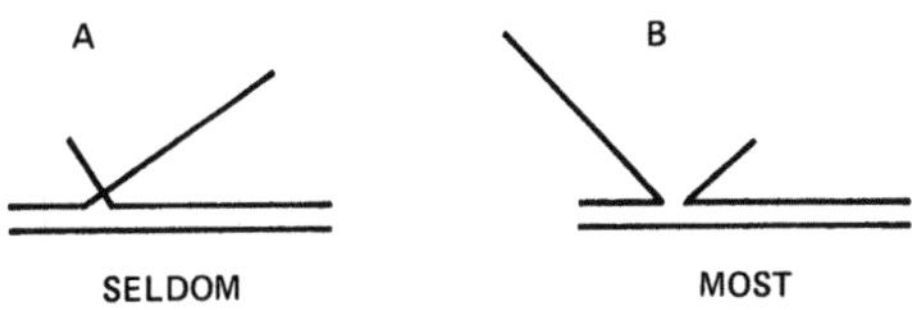

Fig. 18. A and B. Single strands at 2 : 2 branches in DNA. Usually more than 90% of the 2 : 2 branches have the structure shown in B

orientation purposes. The left end of the molecule shown in Figure 18 B can, in general (but not absolutely), be identified if one of the two single strands is of unequal length. If the left end of the duplex molecule is identifiable by some other marker, then the two single strands can be distinguished from one another. This could be important if one were mapping a structure (such as an RNA, protein, or secondary structure feature) on one of the single strands.

VI. Renaturation of Single-Stranded Circular Molecules

If one were to consider the problems associated with the renaturation of two complementary single-stranded circular molecules in solution, one would probably conclude that since the two molecules are prevented from forming a Watson-Crick helix, they would not be able to completely renature. When the results of such an attempted renaturation are viewed in the electron microscope, a class of unique structures can be identified (Fig. 19). The two single-stranded circles are found associated with one another in a partially single-stranded, partially duplex structure, frequently having a few superhelical turns. The two single strands in the denatured regions are generally of equal length. Therefore, the duplex regions are probably between complementary sequences. It seems possible that there are duplex regions of both right- and left-handed helices.

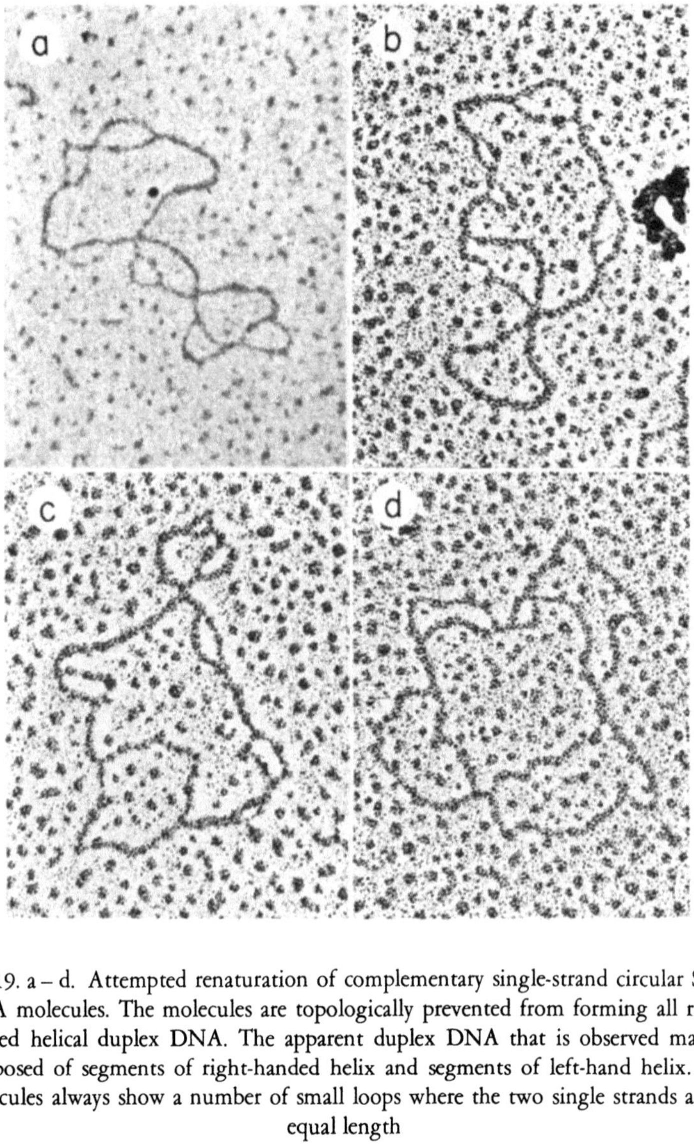

Fig. 19. a – d. Attempted renaturation of complementary single-strand circular SV40 DNA molecules. The molecules are topologically prevented from forming all right-handed helical duplex DNA. The apparent duplex DNA that is observed may be composed of segments of right-handed helix and segments of left-hand helix. The molecules always show a number of small loops where the two single strands are of equal length

VII. Topologic Restriction to Renaturation in Linear Molecules — Renaturation of 'Knotted' DNA

The first step that must occur during the renaturation of DNA to form a heteroduplex molecule is a nucleation event somewhere along the two strands, usually near the ends of the molecules (due to excluded volume effects, DAVIS et al., 1971). The ensuing zippering process occurs very rapidly until a region of nonhomology is encountered, where a new nucleation event is required past the nonhomology region.

This second nucleation may occur near the other end, followed by zippering toward the first renatured region. If there are any topologic constraints in the molecule, such as knots in one of the strands or between the two strands, or cross links (covalent or hydrogen bonding) between two points on the same strand, then this second nucleation event can 'freeze in' the topologic constraint. Such molecules behave as though there is a closed single-strand circular molecule within a linear molecule in that the molecule cannot fully renature. What appears to be such a molecule is shown in Figure 20. The molecule contains a large tangled region in the center, similar in ap-

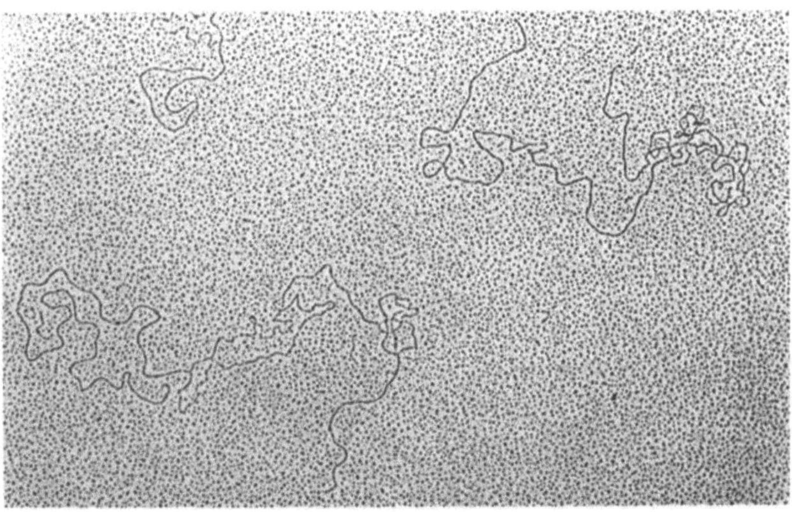

Fig. 20. Two renatured λ heteroduplex molecules. The molecule in the lower left has renatured completely and can be followed easily. The molecule in the upper right has not renatured fully. The central region of this latter molecule appears similar to the renatured circular molecules shown in Figure 19. This molecule may be topologically prevented from full renaturation

pearance to renatured single-strand circular molecules. These molecules are not an artifact of the mounting procedure since a pure preparation of such molecules has been made. These molecules, when covalently closed, are found at a density between covalently closed pure duplex DNA and open circular DNA in cesium chloride-ethidium bromide equilibrium gradients.

F. RNA and Transcription Complexes

The visualization of RNA in the electron microscope has proved to be more difficult than that of DNA. In general, the techniques that have been used have been less easily applied, and are less reproducible, than is the basic protein film technique for visualizing DNA. RNA has stronger intramolecular interactions, and thus more secondary structure, than does either double- or single-stranded DNA. The presence of secondary structure, which may or may not be visible in the electron microscope, depending on its extent, makes obtaining reliable length measurements difficult. Various techniques have been developed with the aim of destabilizing the intramolecular interactions in RNA, ranging from the use of chemical denaturants such as DMSO, urea, and formamide, to specific chemical reactions of the RNA with reagents such as glyoxal or DNA unwinding protein to mask reactive groups in the RNA. A complicating factor in using length measurements of RNA to derive quantitative molecular weight information is the fact that the average spacing of bases in the polynucleotide chain apparently varies over a large range for different classes of RNA and different conditions. The spacing can vary from 2.5 Å in closest stacking to about 6.0 Å in a fully extended form (GRANBOULAN and SCHERRER, 1969). It is not clear whether this variation results from the inherent structure of the RNA in solution, from residual secondary structure, or is an artifact resulting from the preparation procedures. In contrast, the spacing for DNA in solution and on an electron microscope grid is relatively well defined.

I. Techniques for Preparing RNA

Several techniques have been used to prepare samples of RNA that are satisfactory for obtaining length measurements (GRANBOULAN and SCHERRER, 1969; VERMA et al., 1970; ROBBERSON et al., 1971; NANNINGA et al., 1972 b; HSU et al., 1973). These techniques yield length distributions that typically have a standard deviation comparable to, or somewhat greater than, the standard deviation of an equivalent population of DNA molecules.

GRANBOULAN and SCHERRER (1969) used an aqueous-urea method of preparing RNA to measure the lengths of ribosomal, preribosomal, and messenger-like RNA from various animal cells. They spread a solution containing 1 – 4 μg/ml of RNA, 4 – 5 M urea, and .01% chymotrypsinogen or cytochrome c, onto a hypophase of .015 M ammonium acetate (pH 8.0). From the measured lengths they calculated the molecular weights of the various RNA species, using an interbase spacing value of 3.17 Å for messenger-like RNA, and 2.45 Å for ribosomal RNA. These values are based on the measured length of an RNA species for which the number of nucleotides has been chemically determined, R17 phage RNA in the case of messenger-like RNA and *E. coli* 16 and 23S RNA in the case of ribosomal RNA. These interbase spacing values were found to give molecular weights for the various species of RNA that correlate well with molecular weights previously determined by other methods.

NANNINGA et al. (1972 b) prepared *E. coli* ribosomal RNA for electron microscopy without the use of a basic protein film, by dissolving the RNA in 80% DMSO. A drop of the RNA solution is placed on the grid and the excess blotted off with filter paper. The standard deviation of the length distribution determined by this technique is apparently somewhat greater than that determined by several of the other techniques.

ROBBERSON et al. (1971) determined the size of the HeLa mitochondrial specific ribosomal RNA components by using a urea-formamide method. The RNA is spread at a concentration of 0.20 to 0.25 μg/ml from a solution containing 50 μg/ml cytochrome c, 4 M urea, .04 M Tris, .001 M EDTA, and 80% (w/v) formamide, onto a hypophase of 10^{-6} M Tris (pH 7.8), 2.5×10^{-7} M EDTA, and 1.4×10^{-3}% diethylpyrocarbonate. The molecular weight values they calculate are in better agreement with the relative values predicted from sedimentation velocity measurements than the relative values based on gel electrophoresis.

HSU et al. (1973) chemically reacted Sindbis virus RNA with glyoxal to block some of the hydrogen-bonding functions and thus increased the ease

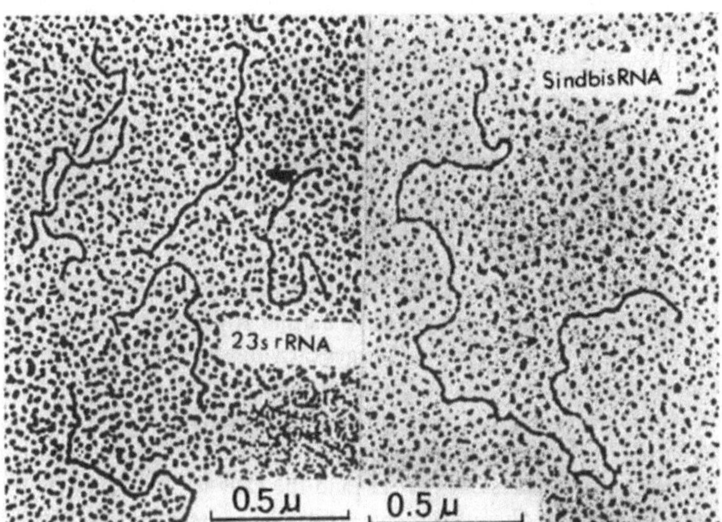

Fig. 21. Sindbis RNA and *E. coli* 23S RNA after treatment with glyoxal. This micrograph was kindly provided by Dr. MING-TA HSU

with which the RNA molecule can be extended. The RNA is heated for 1 h with 0.5 M glyoxal at 37° C in .01 M phosphate buffer (pH 7) and then diluted by a factor of about 100 into the spreading solution and spread from 40 or 50% formamide by the standard basic protein film procedure. Under these conditions probably most of the G residues, and only the G residues, are present as glyoxal adducts. The RNA molecules appear as extended thin threads (Fig. 21), and length distributions with reasonable standard deviations are obtained. Upon spreading the RNA after varying extents of reactions with glyoxal, circular molecules with duplex 'handles' or 'stems' were observed. Thus it was determined that Sindbis virus RNA has cohesive ends. Hybridization of Sindbis virus RNA with poly(dT) and visualization of Sindbis RNA-poly(dT) heteroduplex molecules indicated the existence of a poly(A) sequence at or near one end of the Sindbis RNA molecule.

II. Secondary Structure Maps

The secondary structure in RNA has been used to advantage to locate and orient the 18 and 28S HeLa ribosomal RNA within the 45S RNA precursor by WELLAUER and DAWID (1973). They made use of the observation that HeLa ribosomal RNA molecules have a characteristic and reproducible secondary structure, which is apparently due to the presence of hairpin loops located along the length of the molecule. The RNA was prepared for electron microscopy by the urea-formamide technique of ROBBERSON et al. (1971). The regions of secondary structure were mapped within the ribosomal and ribosomal precursor RNAs. A comparison of the secondary structure maps for these various classes of RNA molecules allowed the positioning and orientation of the smaller molecules within their precursor molecules. Based on these data a processing pathway was postulated that is in good agreement with earlier work.

Secondary structure maps can also be made with DNA (WELLAUER et al., 1974). WU and DAVIDSON (1973 a) described small secondary hairpin-type structures in the single-strand region of heteroduplexes of λ h 80 and λ h 80 transducing phages carrying three transfer RNA genes and also in heteroduplexes of Φ 80 h transducing phages carrying single or duplicate copies of a tyrosine tRNA gene. Figure 22 shows the considerable secondary structure that is found in the immunity region of bacteriophage λ DNA. The secondary structure is found in the genetically undefined region between the terminal end of the repressor gene (cI) and the early left promoter (P_L), left operator (O_L) region. Presumably this structure results from the occurrence of a short inverted repeated sequence. The sequence is not totally homologous because it has a depressed melting temperature (about 35% formamide under standard mounting conditions).

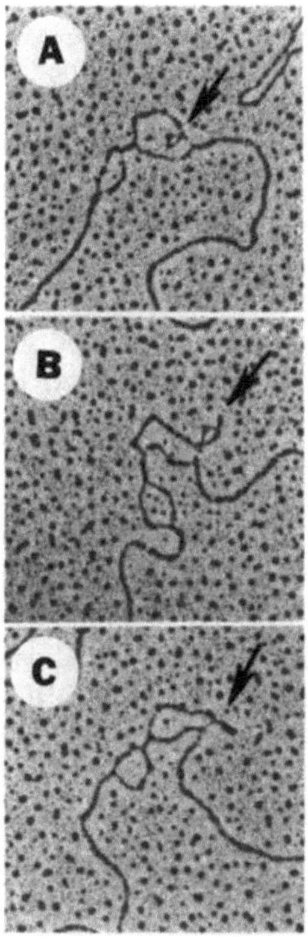

Fig. 22. A – C. The immunity region of three $\lambda/\lambda imm434$ heteroduplex molecules. In A a small loop is seen *(arrow)* in the λ strand of the nonhomology region. In B the loop appears partially duplex *(arrow)*. In C the loop appears fully duplex *(arrow)*. This structure is in the left side of the λ immunity region and just to the right of the left operator

III. Transcription Complexes

Transcription complexes were first visualized by MILLER and BEATTY (1969 a, b), who obtained the now classic micrographs of the active transcription of the ribosomal RNA regions of the nuclear contents of developing amphibian oocytes. Transcription complexes may be visualized by the standard aqueous basic protein film technique. DAVIS and HYMAN (1970) studied the in vitro transcription of T 7 DNA by *E. coli* RNA polymerase. In this case the RNA is not extended and may not be measured, although its size can be qualitatively estimated from the size of the collapsed bush. The minimum size for an observable bush is about 400 nucleotides. The RNA polymerase is not seen because the DNA molecule is surrounded by a layer of denatured cytochrome c molecules. By measuring the position of the RNA bushes on the DNA as a function of time, it is possible to obtain information about the positions of transcription initiation and termination sites, the direction and average rate of transcription, and the average number of RNA polymerase molecules actively transcribing the DNA. The position of a single unique in vitro initiation region for *E. coli* RNA polymerase on T7 DNA was mapped. In addition, one can study the effects of additional factors, such as the transcription termination factor ρ.

DELIUS et al. (1973) described a method for visualizing transcription complexes of T7 and SV40 DNA with *E. coli* RNA polymerase, with the RNA in extended form. They bound bacteriophage T4 gene 32 protein to the RNA and, following fixation with glutaraldehyde, prepared the transcription complexes for electron microscopy by the standard basic protein film technique, spreading from a solution containing 30% formamide. In this way one can determine both the position and the length of the RNA chains. Using this technique they were able to demonstrate the presence of a second low efficiency promotor in T7 DNA in addition to the promoter region mapped by DAVIS and HYMAN (1970). They also demonstrated that transcription of SV40 DNA by *E. coli* RNA polymerase in vitro can initiate at several different locations. Length measurements of the RNA indicate that a single RNA polymerase molecule can transcribe more than 4 times the length of one SV40 DNA strand, passing its own initiation site.

IV. Mapping of Complementary RNA Sequences in DNA

1. R-Loop Method

The term R-loop is used to refer to the structures that are observed in the electron microscope when RNA is hybridized to duplex DNA that contains sequences complementary to the RNA. Upon hybridization of the RNA strand to its complementary DNA strand with the concomitant displacement of the identical DNA strand, a loop consisting of an RNA/DNA hybrid strand and of a single strand of DNA is formed (Fig. 23). R-loops are identical in appearance of the D-loops reported by ROBBERSON et al. (1972). WHITE and HOGNESS (1977) observed that rRNA can hybridize to duplex rDNA from Drosophila melanogaster.

THOMAS et al. (1976) studied the rate of R-loop formation and the kinetic stability of R-loops under a variety of conditions. They used as a simple model system a viable λ hybrid DNA molecule containing the 2.5 kilobase pair repetitive *Eco*RI DNA fragment from *Saccharomyces cerevisiae* (yeast) in-

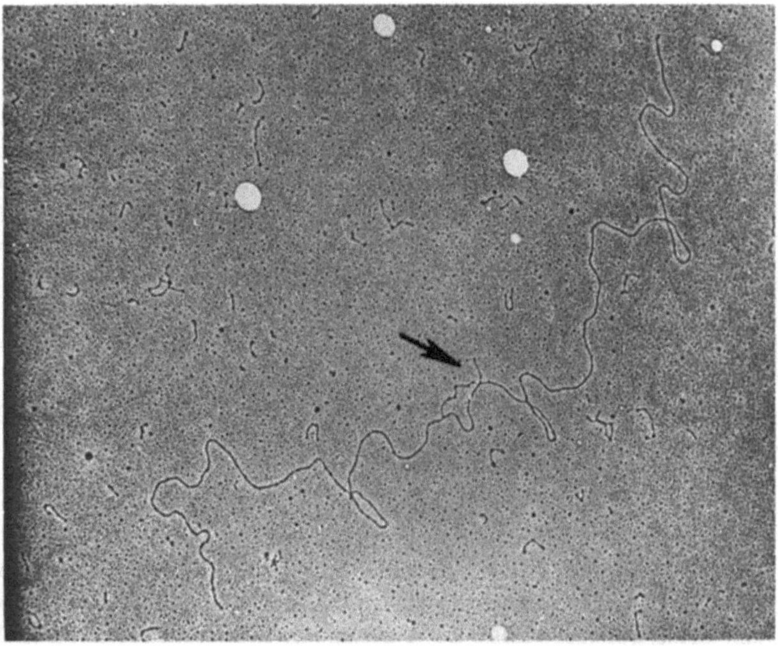

Fig. 23. A λ DNA molecule containing an R-loop (*arrow*). The λ DNA molecule contains a yeast DNA fragment that hybridizes in yeast 26S RNA. The short single-strand tail is one end of the yeast 26S RNA molecule. This micrograph was kindly provided by MARJORIE THOMAS

serted into the middle of bacteriophage λ DNA. The inserted yeast DNA fragment hybridizes to yeast 26 S rRNA.

R-loop reaction mixtures contained 50 μl of a buffered formamide solution and 10 μl of a solution containing 60 μg/ml each of RNA and DNA, 0.1 M NaCl, and 0.05 M Tris (pH 7.5). The buffered formamide solution was prepared by mixing 0.42 ml crystallized formamide, 50 μl 1 M PIPES (pH 7.8), 12 μl 0.5 M Na_3EDTA, and 18 μl H_2O.

They found that the rate of R-loop formation is very dependent on temperature; the maximal rate is comparable to the rate of hybridization of RNA to single-stranded DNA and occurs within one degree of the denaturation temperature of the DNA, falling precipitously a few degrees above or below that temperature. Unter appropriate conditions it is possible to saturate the DNA with R-loops.

R-loop formation in formamide occurs because of the increased thermodynamic stability of the RNA/DNA hybrid relative to the DNA/DNA hybrid near the denaturation temperature of duplex DNA. After the formamide is removed, R-loops are not thermodynamically stable and should be displaced by branch migration. However, THOMAS et al. (1976) have observed that in the absence of formamide, R-loops have remarkable kinetic stability, and destruction of the RNA by alkali or RNase treatment seems necessary for their elimination. R-loops are quantitatively retained after storage at 5° C for substantial lengths of time either in the formamide reaction mixture or after removal of the formamide by dialysis.

Since R-loops are kinetically stable after the removal of the formamide, it is possible to cleave DNA molecules containing R-loops with restriction endonucleases, as was demonstrated by THOMAS et al. with the *Eco*RI endonuclease. This makes possible fine structure mapping relative to restriction endonuclease cleavage sites.

The R-loop method should prove to be an extremely powerful mapping tool for any coding sequence in DNA for which an RNA can be obtained.

2. *Single-Strand Binding Protein Method*

Coding sequences in DNA have been mapped by the direct observation of RNA/DNA duplex regions along a DNA single strand in several cases where the length of the RNA/DNA duplex region is on the order of several kilobase pairs. OHTSUBO et al. (1974) mapped the positions of the 16S and 23S genes on the DNA of the transducing phage Φ 80 d_3*ilv*$^+$ *su*$^+$ 7 by direct observation of the rRNA/DNA hybrid regions. HYMAN (1971) mapped the early coding region of bacteriophage T7 DNA by direct observation of the hybrid region formed by the four early messenger RNAs on T7 DNA. He found that they all mapped contiguously over a region of about 6 kilobase pairs.

It becomes very difficult, due to insufficient contrast, to visualize directly an RNA/DNA duplex region of much less than several hundred base pairs in length along a DNA single strand. WU and DAVIDSON (1975) developed a method for mapping sequences in DNA complementary to very small RNAs, by using a single-strand binding protein to enhance the contrast between the single strand and RNA/DNA duplex regions. The distinction was further enhanced by mounting the DNA in the presence of ethidium bromide on carbon-coated Parlodion grids (see Sect. H II). Unter these conditions, the thickness of the duplex regions is further reduced relative to that of the protein-coated single strands. They have used this method to map the positions of the 16S, 23S, and 5S rRNA genes and the several tRNA genes present on the DNA of the transducing phage Φ 80 d_3 *ilv*$^+$ *su*$^+$ 7. RNA/DNA hybridizations were carried out at 53° for 5 min in 50% formamide, 0.1 M Tris (pH 8.5), 0.01 M Na_3EDTA, 0.65 M NaCl. The DNA concentration was 1 μg/ml, and concentrations of various RNA species ranged from 5 to 250 μg/ml. This reaction mixture was dialyzed into 0.01 M phosphate buffer (pH 7.4), 0.001 M Na_2EDTA. The single-strand binding protein reaction was carried out with 20 to 50 μg/ml gene 32 protein and 0.5 to 1 μg/ml DNA at 37° for 5 min, followed by fixation with 0.2% glutaraldehyde.

G. Tagging Methods

There is a limit to the size and type of structure that it is possible to visualize using the standard aqueous and formamide basic protein techniques that have been discussed so far in this Chapter. For instance, it is not possible, due to the problem of insufficient contrast, to visualize along a single-stranded DNA molecule a DNA/DNA or DNA/RNA duplex structure of much less than several hundred base pairs in length. And, because of the intrinsic nature of the basic protein technique, it is not possible to look at complexes of proteins bound to DNA. The cytochrome c that collapses around the DNA molecule obscures from visibility any proteins that are previously bound. In some cases it has been possible to circumvent these problems by specifically attaching to a small molecule of interest a structure that is visible in the electron microscope. The 'tag' that has so far been used successfully is the molecule ferritin, which because of its electron-dense iron core is opaque to the electron beam and therefore readily recognizable in the electron microscope. This technique has been used to map the binding positions of two classes of small RNAs, and of one protein, to DNA. The 4S RNA genes of HeLa mitochondrial DNA (WU et al., 1972) and the *E. coli* tRNA genes carried by several transducing phages (WU and DAVIDSON, 1973 b; WU et al., 1973) have been mapped using ferritin-conjugated RNA species. The position of binding of SV40 T antigen to SV40 DNA has been mapped with the use of ferritin-conjugated antibodies (REED et al., 1975).

I. RNA-Ferritin Tags

WU and DAVIDSON (1973 b) studied in detail three different schemes for coupling RNA to ferritin. The coupling was most successfully accomplished after chemical activation of the ferritin by acylation with bromoacetic anhydride. The number of acyl groups per ferritin molecule as a function of the amount of bromoacetic anhydride in the reaction was determined, as well as the efficiency of coupling of the ferritin to the RNA as a function of the number of acyl groups per ferritin molecule. The acylated ferritin can be separated from native ferritin and fractionated according to its number of acyl groups by salt elution from DEAE-cellulose. Thus the coupling step can be optimized by fractionation of the acylated ferritin. The RNA was modified prior to the ferritin coupling by oxidation of the 3′ end to the dialdehyde with periodate, and then reaction either with cystamine followed by reduction with sodium borohydride and dithiothreitol, or with meracaptosuccinic dihydrazide. The coupling step is accomplished by reaction of the carbon-bromine bond of the activated ferritin and the sulfhydryl bond of the modified RNA. As ferritin sediments between 50 and 70S and the small RNA species used sediment at 4S, it is possible to separate the ferritin-RNA complex from the uncoupled RNA on a sucrose gradient. This is necessary because hybridizes more easily to the DNA than does the ferritin-RNA complex. Thus, the presence of any uncoupled RNA would compete out the binding of the ferritin-RNA complex. To obtain effective hybridization at a higher concentration of ferritin-RNA complex it is desirable to separate the uncoupled ferritin by DEAE-cellulose chromatography. After the hybridization is complete, the presence of the excess ferritin-RNA complex causes a high background of ferritins in the electron microscope unless it is first removed. This has been accomplished in two ways, either by a gentle phenol extraction or a gel electrophoresis step. An example of the kind of DNA molecule with ferritins bound that it is possible to see is shown in Figure 24.

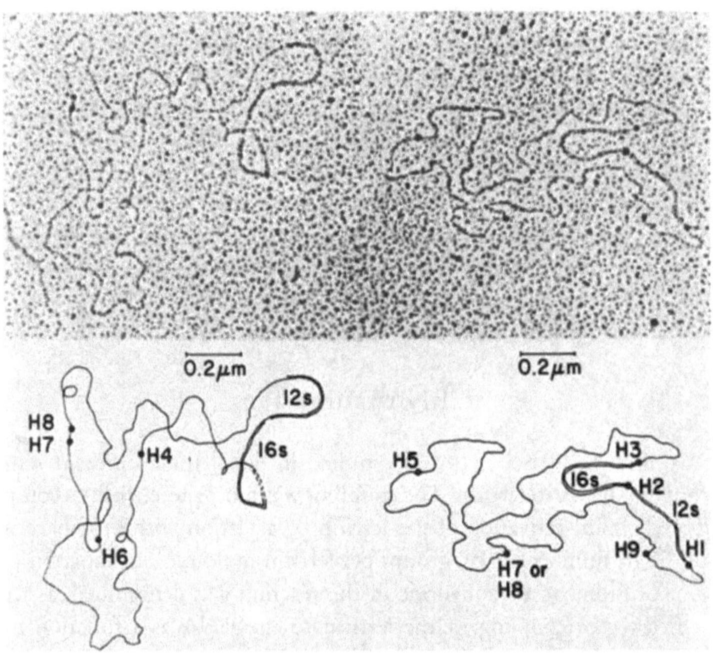

Fig. 24. HeLa mitochondrial H-strand DNA molecules annealed to 12S RNA, 16S RNA, and 4S RNA-ferritin conjugates. The 4S RNA-ferritin binding sites have been numbered H1 to H9. Explanatory drawings of each molecule are shown below the micrographs. This figure was kindly provided by Dr. NORMAN DAVIDSON

II. Protein-Ferritin Tags

REED et al. (1975) used the ferritin-tagging method to study the binding of SV40 T antigen to SV40 DNA. In this case an indirect antibody method was used. SV40 T antigen was bound to SV40 DNA and fixed with glutaraldehyde. Anti-T antibody obtained from hamsters was bound to the SV40 DNA-T antigen complex, followed by ferritin-conjugated goat antihamster antibody. The whole complex was fixed again with glutaraldehyde. The ferritin-antibody coupling was accomplished by prior activation of the ferritin with glutaraldehyde (OLSEN et al., 1973). The coupling was carried out in antibody excess, and the ferritin-antibody complex separated from uncoupled antibody by gel filtration. An example of an SV40 DNA molecule with a ferritin bound is shown in Figure 25.

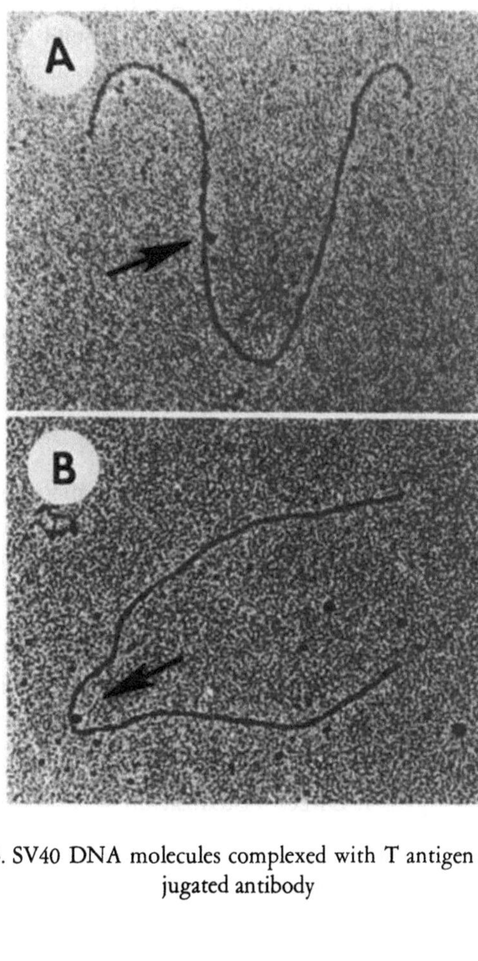

Fig. 25. A and B. SV40 DNA molecules complexed with T antigen and ferritin-conjugated antibody

III. General Comments and Problems

The ferritin-tagging technique is at a relatively early stage of development. Various aspects and problems related to its use are not yet well understood. The hybridization of the ferritin-coupled RNA species to the DNA is apparently much slower than that of the native RNA species; the half-time for saturation is approximately 100 times slower. The reason for this is not known. No kinetic studies have been done of the binding of the ferritin-conjugated antibodies. It is not known whether it is possible to completely saturate all the available binding sites with either the ferritin-conjugated RNA or antibody. Due to the high concentration of ferritin-conjugated RNA needed for hybridization with the DNA, the background of ferritin on the grid is a significant problem unless the excess ferritin-coupled RNA is removed from the DNA after hybridization. The background of ferritin on the grid was not found to be a problem with the ferritin-conjugated antibody, as lower concentrations of the ferritin reagent could be used. REED et al. (1975) observed that the fraction of DNA molecules with ferritin bound that sticks to the grid is apparently variable. The fraction of DNA molecules with ferritin bound varied between 1 and 15% in different preparations of the same sample, and there was similar variation from one part of an electron microscope grid to another. This phenomenon makes obtaining data about the frequency of molecules with bound ferritin problematical in the absence of independent data on the fraction of DNA molecules with ferritin bound in solution.

The ferritin-tagging technique should prove to be valuable for probing specific interactions between RNAs, proteins, and DNA in the future, despite the present problems. DAVIDSON and his co workers are continuing to explore other possible tagging methods that may prove to be useful as well.

H. Protein-free Spreading

Because the DNA molecules seen in the electron microscope after preparation by the basic protein film technique are thickly coated with cytochrome c, the use of this technique precludes the possibility of examining protein-nucleic acid complexes. Numerous attempts have been made to develop protein-free spreading techniques for the purpose of visualizing, and doing high resolution microscopy of, protein-nucleic acid interactions. These attempts have achieved varying degrees of success and have not as yet become widely used; however, several of these techniques appear to hold promise for future widespread use.

I. Direct Visualization

GRIFFITH (1973) described a method for the direct visualization of bare DNA and DNA-protein complexes that involves adsorption of the DNA to a glow-activated thin carbon grid, dehydration in ethanol, and contrast enhancement by vacuum deposition of a very small amount of tungsten. This technique has been used to examine a variety of DNA-protein complexes,

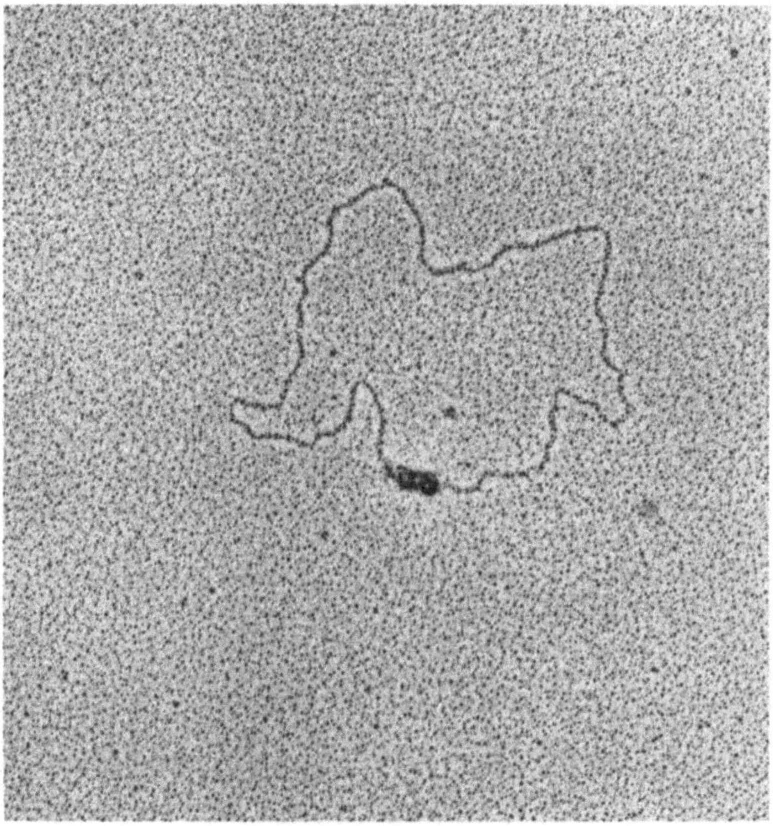

Fig. 26. Direct visualization of SV40 DNA with a protein moiety bound. This micrograph was kindly provided by Dr. JACK GRIFFITH

including those with *E. coli* DNA polymerase I, lambda repressor, histones, and for the examination of eukaryotic chromatin fibers. A complex of protein and SV40 DNA obtained from SV40-infected monkey cells or mature SV40 virions has been detected and mapped by this method (GRIFFITH et al., 1975). This complex (Fig. 26) consists of a circular SV40 DNA molecule with a protein moiety bound at a location near the origin of DNA replication.

II. Intercalating Dye Method

A method for the visualization of double-stranded DNA employing intercalating dyes to mediate the adsorption of the DNA to carbon or mica films was described by KOLLER et al. (1974) and PORTMANN et al. (1974). The dyes ethidium bromide, propidium diiodide, and actinomine have all been used successfully. The DNA in 50 to 250 μg/ml of dye is adsorbed onto a freshly cleaved mica sheet. After washing and drying the mica, a carbon replica is made, picked up on a grid, and shadowed with platinum. This method yields length measurements of double-stranded DNA with standard deviations comparable to those obtained from length measurements of DNA prepared by the basic protein film method, and has been used to look at complexes of RNA polymerase with T3 and T7 DNA. Histograms with sharp peaks indicating specific binding positions have been obtained.

III. Benzyldimethylalkylammonium Chloride Method

The third method that looks promising for possible widespread future use has been described by VOLLENWEIDER et al. (1975). This method is very similar to the basic protein film technique, but uses the compound benzyldimethylalkylammonium chloride in place of cytochrome c. Benzyldimethylalkylammonium chloride is very much smaller than cytochrome c, yielding double-stranded DNA molecules with about a third the diameter of DNA molecules spread with cytochrome c. With this method it is also possible to prepare extended single-stranded DNA. The buffered DNA solution contains $2.5 \times 10^{-3}\%$ benzyldimethylalkylammonium chloride and 1.3% formamide. This solution can either be spread as has been described before for the basic protein film method, or the DNA can be adsorbed directly onto carbon-coated grids by soaking the grid in the DNA solution. In either case the grid is subsequently washed in water, dehydrated in ethanol, dried, and platinum shadowed. This method appears to give length distributions for both single- and double-stranded DNA with standard deviations that are comparable to those obtained with the basic protein film method. It is also possible to visualize RNA polymerase bound to T7 DNA with this method.

We have used this method to map a small deletion of about 150 base pairs in SV40 DNA (unpublished results). We have observed that lateral aggregation of duplex DNA, and insufficient contrast of single-strand DNA, can be problems in using this method. Lateral aggregation of duplex strands is more severe with longer DNAs than with shorter DNAs and is alleviated somewhat by spreading onto a hypophase of distilled H_2O.

IV. Other Methods

Several other methods for protein-free spreading and high resolution microscopy have been reported in the literature are uranyl formate staining (BRACK and PIRROTTA, 1975), a two-step method of mounting DNA on cytochrome c monolayers and shadowing with tantalum-tungsten (ABERMANN and SALPETER, 1974), Brij-58 preparation of RNA (THACH and THACH, 1973), 80% DMSO preparation of RNA and RNA-protein complexes (NANNINGA et al., 1972 a, b), the use of amylamine vapor to charge carbon grids (DUBOCHET et al., 1971; BORDIER and DUBOCHET, 1974), and the use of dark-field electron microscopy (DUBOCHET et al., 1971; BORDIER and DUBOCHET, 1974).

References

Abermann, R., Salpeter, M. M.: Visualization of deoxyribonucleic acid molecules by protein film adsorption and tantalum-tungsten shadowing. J. Histochem. Cytochem. **22**, 845 – 855 (1974)

Bigger, C. H., Murray, K., Murray, N. E.: Recognition sequence of a restriction enzyme. Nature (New Biol.) **244**, 7 – 10 (1973)

Bordier, C., Dubochet, J.: Electron microscopic localization of the binding sites of *Escherichia coli* RNA polymerase in the early promoter region of T7 DNA. Europ. J. Biochem. **44**, 617 – 624 (1974)

Brack, C., Pirrotta, V.: Electron microscopic study of the repressor of phage λ and its interaction with operator DNA. J. Mol. Biol. **96**, 139 – 152 (1975)

Brockman, W. W., Lee, T. N. H., Nathans, D.: Characterization of cloned evolutionary variants of simian virus 40. Cold Spring Harbor Symp. Quant. Biol. **39**, 119 – 127 (1974)

Broker, T. R., Lehman, I. R.: Branched DNA molecules: Intermediates in T4 recombination. J. Mol. Biol. **60**, 131 – 149 (1971)

Cassuto, E., Radding, C. M.: Mechanism for the action of λ exonuclease in genetic recombination. Nature (New Biol.) **229**, 13 – 16; **230**, 128 (1971)

Chow, L. T., Scott, J. M., Broker, T. R.: The electron microscopy of nucleic acids — laboratory manual. Cold Spring Harbor, New York: Cold Spring Harbor Laboratory 1975

Cuzin, F., Vogt, M., Dieckmann, M., Berg, P.: Induction of virus multiplication in 3T3 cells transformed by a thermosensitive mutant of polyoma virus. J. Mol. Biol. **47**, 317 – 333 (1970)

Davis, R. W., Davidson, N.: Electron microscopic visualization of deletion mutations. Proc. Natl. Acad. Sci. U. S. **60**, 243 – 250 (1968)

Davis, R. W., Hyman, R. W.: Physical locations of the *in vitro* RNA initiation site and termination sites of T7M DNA. Cold Spring Harbor Symp. Quant. Biol. **35**, 269 – 281 (1970)

Davis, R. W., Hyman, R. W.: A study in evolution: The DNA base sequence homology between coliphages T7 and T3. J. Mol. Biol. **62**, 287 – 301 (1971)

Davis, R. W., Simon, M., Davidson, N.: Electron microscope heteroduplex methods for mapping regions of base sequence homology in nucleic acids. In: Methods in Enzymology. Grossman, L., Moldave, K. (eds.). New York-London: Academic Press, 1971, Vol. XXI, pp. 413 – 428

Delius, H., Westphal, H., Axelrod, N.: Length measurements of RNA synthesized *in vitro* by *Escherichia coli* RNA polymerase. J. Mol. Biol. **74**, 677 – 687 (1973)

Dubochet, J., Ducommun, M., Zollinger, M., Kellenberger, E.: A new preparation method for dark-field electron microscopy of biomacromolecules. J. Ultrastruct. Res. **35**, 147 – 167 (1971)

Ferguson, J., Davis, R. W.: An electron microscopic method for studying and mapping the region of weak sequence homology between simian virus 40 and polyoma DNAs. J. Mol. Biol. **94**, 135 – 149 (1975)

Granboulan, N., Scherrer, K.: Visualization in the electron microscope and size of RNA from animal cells. Europ. J. Biochem. **9**, 1 – 20 (1969)

Greenfield, L., Simpson, L., Kaplan, D.: Conversion of closed circular DNA molecules to single-nicked molecules by digestion with DNAase I in the presence of ethidium bromide. Biochim. Biophys. Acta **407**, 365 – 375 (1975)

Griffith, J. D.: Electron microscopic visualization of DNA in association with cellular components. In: Methods in Cell Biology. Prescott, D. M. (ed.). New York-London: Academic Press, 1973, Vol. VII, pp. 129 – 146

Griffith, J., Dieckmann, M., Berg, P.: Electron microscope localization of a protein bound near the origin of simian virus 40 DNA replication. J. Virol. **15**, 167 – 172 (1975)

Hedgpeth, J., Goodman, H. M., Boyer, H. W.: DNA nucleotide sequence restricted by the RI endonuclease. Proc. Natl. Acad. Sci. U. S. **69**, 3448 – 3452 (1972)

Hsu, M.-T., Kung, H.-J., Davidson, N.: An electron microscope study of Sindbis virus RNA. Cold Spring Harbor Symp. Quant. Biol. **38**, 943 – 950 (1973)

Hudson, B., Upholt, W. B., Devinny, J., Vinograd, J.: The use of an ethidium analogue in the dye-buoyant density procedure for the isolation of closed circular DNA: the Variation of the superhelix density of mitochondrial DNA. Proc. Natl. Acad. Sci. U. S. **62**, 813 – 820 (1969)

Hyman, R. W.: Physical mapping of T7 messenger RNA. J. Mol. Biol. **61**, 369 – 376 (1971)

Inman, R. B.: Denaturation maps of the left and right sides of the lambda DNA molecule determined by electron microscopy. J. Mol. Biol. **28**, 103 – 116 (1967)

Inman, R. B., Schnös, M.: Partial denaturation of thymine- and 5-bromouracil-containing lambda DNA in alkali. J. Mol. Biol. **49**, 93 – 98 (1970)

Kelly, T. J., Jr., Smith, H. O.: A restriction enzyme from *Hemophilus influenzae* II. Base sequence of the recognition site. J. Mol. Biol. **51**, 393 – 409 (1970)

Kim, J.-S., Sharp, P. A., Davidson, N.: Electron microscopic studies of heteroduplex DNA from a deletion mutant of bacteriophage ΦX-174. Proc. Natl. Acad. Sci. U. S. **69**, 1948 – 1952 (1972)

Kleinschmidt, A. K.: Monolayer techniques in electron microscopy of nucleic acid molecules. In: Methods in Enzymology. Colowick, S. P., Kaplan, N. O. (eds.). New York-London: Academic Press, 1968, Vol. XII B, pp. 361 – 377

Kleinschmidt, A. K., Zahn, R. K.: Über Desoxyribonucleinsäure-Molekeln in Protein-Mischfilmen. Z. Naturforsch. B **14**, 770 – 775 (1959)

Koller, T., Sogo, J. M., Bujard, H.: An electron microscopic method for studying nucleic acid-protein complexes. Visualization of RNA polymerase bound to the DNA of bacteriophages T7 and T3. Biopolymers **13**, 995 – 1009 (1974)

Lee, C. S., Davis, R. W., Davidson, N.: A physical study by electron microscopy of the terminally repetitious, circularly permuted DNA from the coliphage particles of *Escherichia coli* 15. J. Mol. Biol. **48**, 1 – 22 (1970)

Massie, H. R., Zimm, B. H.: Kinetics of denaturation of DNA. Biopolymers **7**, 475 – 493 (1969)

McConaughy, B. L., Laird, C. D., McCarthy, B. J.: Nucleic acid reassociation in formamide. Biochemistry **8**, 3289 – 3295 (1969)

Mertz, J. E.: Ph. D. Dissertation, Stanford University Library (1975)

Mertz, J. E., Carbon, J., Herzberg, M., Davis, R. W., Berg, P.: Isolation and characterization of individual clones of simian virus 40 mutants containing deletions, and insertions in their DNA. Cold Spring Harbor Symp. Quant. Biol. **39**, 69 – 84 (1974)

Mertz, J. E., Davis, R. W.: Cleavage of DNA by RI restriction endonuclease generates cohesive ends. Proc. Natl. Acad. Sci. U.S. **69**, 3370 – 3374 (1972)

Miller, O. L., Jr., Beatty, B. R.: Visualization of nucleolar genes. Science **164**, 955 – 957 (1969 a)

Miller, O. L., Jr., Beatty, B. R.: Portrait of a gene. J. Cell. Physiol. **74**, Suppl. 1, 225 – 232 (1969 b)

Morrow, J. F., Berg, P.: Cleavage of simian virus 40 DNA at a unique site by a bacterial restriction enzyme. Proc. Natl. Acad. Sci. U. S. **69**, 3365 – 3369 (1972)

Morrow, J. F., Berg, P., Kelly, T. J., Jr., Lewis, A. M., Jr.: Mapping of simian virus 40 early functions on the viral chromosome. J. Virol. **12**, 653 – 658 (1973)

Nanninga, N., Garrett, R. A., Stöffler, G., Klotz, G.: Ribosomal proteins XXXVIII. Electron microscopy of ribosomal protein S4-165 RNA complexes of *Escherichia coli*. Mol. Gen. Genet. **119**, 175 – 184 (1972 a)

Nanninga, N., Meyer, M., Sloof, P., Reijnders, L.: Electron microscopy of *Escherichia coli* ribosomal RNA: Spreading without a basic protein film. J. Mol. Biol. **72**, 807 – 810 (1972 b)

Ohtsubo, E., Soll, L., Deonier, R. C., Lee, H. J., Davidson, N.: Electron microscope heteroduplex studies of sequence relations among plasmids of *E. coli*. J. Mol. Biol. **89**, 631 – 646 (1974)

Olsen, B. R., Berg, R. A., Kishida, Y., Prockop, D. J.: Collagen synthesis: Localization of prolyl hydroxylase in tendon cells detected with ferritin-labeled antibodies. Science **182**, 825 – 827 (1973)

Portmann, R., Sogo, J. M., Koller, T., Zillig, W.: Binding sites of *E. coli* RNA polymerase on T7 DNA as determined by electron microscopy. FEBS Lett. **45**, 64 – 67 (1974)

Raleigh, E. A., Davis, R. W.: Determination of DNA concentration by electron microscopy. Anal. Biochem. **72**, 460 – 467 (1976)

Reed, S. I., Ferguson, J., Davis, R. W., Stark, G. R.: T antigen binds to simian virus 40 DNA at the origin of replication. Proc. Natl. Acad. Sci. U. S. **72**, 1605 – 1609 (1975)

Risser, R., Mulder, C.: Relative locations of rearrangements in the DNA of defective simian virus 40 (SV40). Virology **58**, 424 – 438 (1974)

Robberson, D., Aloni, Y., Attardi, G., Davidson, N.: Expression of the mitochondrial genome in HeLa cells VI. Size determination of mitochondrial ribosomal RNA by electron microscopy. J. Mol. Biol. **60**, 473 – 484 (1971)

Robberson, D. L., Kasamatsu, H., Vinograd, S.: Replication of mitochondrial DNA-Circular replicative intermediates in mouse L cells. Proc. Natl. Acad. Sci. U. S. **69**, 737 – 741 (1972)

Schnös, M., Inman, R. B.: Position of branch points in replicating lambda DNA. J. Mol. Biol. **51**, 61 – 73 (1970)

Studier, W. B.: Effects of the conformation of single-stranded DNA on renaturation and aggregation. J. Mol. Biol. **41**, 199 – 209 (1969)

Tai, H. T., Smith, C. A., Sharp, P. A., Vinograd, J.: Sequence heterogeneity in closed simian virus 40 deoxyribonucleic acid. J. Virol. **9**, 317 – 325 (1972)

Thach, S. S., Thach, R. E.: Mechanism of viral replication I. Structure of replication complexes of R17 bacteriophage. J. Mol. Biol. **81**, 367 – 380 (1973)

Thomas, C. A., MacHattie, L. A.: Circular T2 DNA molecules. Proc. Natl. Acad. Sci. U. S. **52**, 1297 – 1301 (1964)

Thomas, M., Cameron, J. R., Davis, R. W.: Viable molecular hybrids of bacteriophage lambda and eukaryotic DNA. Proc. Natl. Acad. Sci. U. S. **71**, 4579 – 4583 (1974)

Thomas, M., Davis, R. W.: Studies on the cleavage of bacteriophage lambda DNA with *Eco*RI restriction endonuclease. J. Mol. Biol. **91**, 315 – 328 (1974)

Thomas, M., White, R. L., Davis, R. W.: Hybridization of RNA to double stranded DNA: Formation of R-loops. Proc. Natl. Acad. Sci. U. S. **73**, 2294 – 2298 (1976)

Verma, I., Edelman, M., Herzberg, M., Littauer, U.: Size determination of mitochondrial ribosomal RNA from *Aspergillus nidulans* by electron microscopy. J. Mol. Biol. **52**, 137 – 140 (1970)

Vollenweider, H. J., Sogo, J. M., Koller, T.: A routine method for protein-free spreading of double- and single-stranded nucleic acid molecules. Proc. Natl. Acad. Sci. U. S. **72**, 83 – 87 (1975)

Wellauer, P. K., Dawid, I. B.: Secondary structure maps of RNA: Processing of HeLa ribosomal RNA. Proc. Natl. Acad. Sci. U. S. **70**, 2827 – 2831 (1973)

Wellauer, P. K., Reeder, R. H., Carroll, D., Brown, D. D., Deutch, A., Higashinakagawa, T., Dawid, I. B.: Amplified ribosomal DNA from *Xenopus laevis* has heterogeneous spacer lengths. Proc. Natl. Acad. Sci. U.S. **71**, 2823 – 2827 (1974)

Westmoreland, B. C., Szybalski, W., Ris, H.: Mapping of deletions and substitutions in heteroduplex DNA molecules of bacteriophage lambda by electron microscopy. Science **163**, 1343 – 1348 (1969)

Wetmur, J. G., Davidson, N.: Kinetics of renaturation of DNA. J. Mol. Biol. **31**, 349 – 370 (1968)

White, R. L., Hogness, D.: R-loop mapping of the 18S and 28S sequences in the long and short repeating units of *Drosophila melanogaster* rDNA. Cell **10**, 177 – 192 (1977)

Wolfson, J., Dressler, D., Magazin, M.: Bacteriophage T7 DNA replication: A linear replicating intermediate. Proc. Natl. Acad. Sci. U. S. **69**, 499 – 504 (1971)

Wu, M., Davidson, N.: Secondary structure in transfer RNA genes. J. Mol. Biol. **78**, 35 – 41 (1973 a)

Wu, M., Davidson, N.: A Technique for mapping transfer RNA genes by electron microscopy of hybrids of ferritin-labeled transfer RNA and DNA: The Φ 80 hpsu$_{III}^{+,-}$ system. J. Mol. Biol. **78**, 1 – 21 (1973 b)

Wu, M., Davidson, N.: Use of gene 32 protein staining of single-strand polynucleotides for gene mapping by electron microscopy: Application to the Φ 80 d$_3$ *ilv su*$^+$ 7 system. Proc. Natl. Acad. Sci. U. S. **72**, 4506 – 4510 (1975)

Wu, M., Davidson, N., Attardi, G., Aloni, Y.: Expression of the mitochondrial genome in HeLa cells XIV. The relative positions of the 4S RNA genes and of the ribosomal RNA genes in mitochondrial DNA. J. Mol. Biol. **71**, 81 – 93 (1972)

Wu, M., Davidson, N., Carbon, J.: Physical mapping of the transfer RNA genes on λh 80 d gly Tsu$_{36}^+$. J. Mol. Biol. **78**, 23 – 34 (1973)

Younghusband, H. B., Inman, R. B.: The electron microscopy of DNA. Ann. Rev. of Biochem. **43**, 605 – 619 (1974)

Electron Microscopy of Specific Proteins: Three-Dimensional Mapping of Ribosomal Proteins Using Antibody Labels

J. A. LAKE

A. Introduction

Knowledge of ribosome structure and function has increased at a remarkable rate. One of the reasons for this increased understanding has been the very fruitful combination of biochemistry, immunology, and electron microscopy. This Chapter will explain in detail the techniques of electron microscopy and the techniques of interpretation of electron micrographs of antibody-labeled ribosomes and ribosomal subunits in anticipation that these techniques can be usefully extended to the analysis of other cell organelles and supramolecular structures such as microtubules, flagella, nuclear pores, thick and thin muscle filaments, gap junctions, and enzyme complexes.

B. Techniques

The most widely used method of negatively staining macromolecules and macromolecular assemblies is the droplet method described by BRENNER and HORNE (1959). The simplicity of this technique has tended to obscure the important advantages of other negative staining methods. Although the droplet technique is well suited for resolving structural details of many objects, such as viruses, smaller molecules, such as antibodies, cannot be reliably contrasted by this method. It was therefore considered to be a significant advance when VALENTINE and GREEN (1967) used another negative staining technique to resolve reproducibly the shape of IgG molecules. The technique of immune electron microscopy was initially used, in combination with the droplet method, by YANAGIDA and AHMAD-ZADEH (1970) to map the locations of specific proteins in bacteriophage T4. It was first applied to small ribosomal subunits, large ribosomal subunits and monomeric 70S ribosomes by LAKE et al. (1974 a), WABL (1974), and LAKE (1976), respectively. Subsequently, extensive mapping of both small and large subunits has been done in Germany by TISCHENDORF, ZEICHARDT and STÖFFLER and in the United States by KAHAN, STRYCHARZ, NOMURA and LAKE. The technique employed to visualize antibodies (VALENTINE et al., 1968) is still relatively unknown today and is currently referred to as the *single carbon layer* technique. This is in order to distinguish it from a variation of the technique, the *double carbon layer* technique, that is used primarily in antibody-labeling studies of ribosomal subunits.

These two staining methods are illustrated in Figure 1. In both techniques, a small carbon film that has been previously deposited onto the surface of freshly cleaved mica is partially floated onto a solution of ribosomes (or other macromolecules) being stained, as shown in Figure 1 a. The carbon film is then withdrawn by lifting the mica from the solution, and the attached layer of carbon traps some of the ribosome solution between it and the mica, as shown in Figure 1 b. In the second stage of the process (Fig. 1 c), the mica is inserted into a solution of uranyl acetate so that the carbon film is comple-

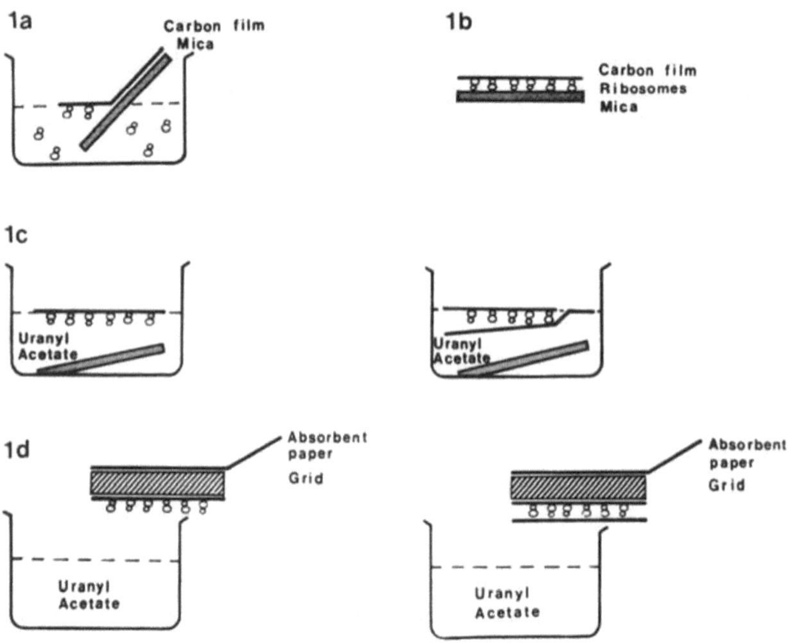

Fig. 1 a – d. A diagram of the technique of preparing single carbon layer and double carbon layer negatively stained specimen grids: (a) adsorbing the ribosomes to the carbon film; (b) the carbon film after withdrawing the mica; (c) floating the carbon film off the mica and onto the surface of the negative stain. The single carbon layer method is illustrated *at the left* and the double carbon layer technique *at the right.* (d) Lifting the grid from the negative stain; completed grids: a single carbon layer grid is shown *at the left* and a double carbon layer grid *at the right*

tely floated off the mica substrate. In the double carbon layer technique (shown at the right in Fig. 1 c), a few small pieces of carbon film have been previously floated onto the surface of the negative staining solution so that when the carbon-coated mica is inserted, the ribosomes are sandwiched in a shell of negative stain between two carbon films. In both methods a grid [1] is placed on top of the carbon and then the grid and attached carbon film(s) are lifted from the uranyl acetate solution by placing a piece of absorbent paper [2] on top of the grid, waiting for the staining solution to begin wetting the absorbent paper and lifting the grid from the solution, as shown in Figure 1 d.

A completed grid is diagrammed in Figure 1 e. A specimen prepared using the double carbon technique should contain regions of the specimens that are negatively stained and sandwiched between two carbon films. Specimens prepared using the single carbon layer technique should contain regions of negatively stained macromolecules upon a single carbon film. Naturally, neither technique will produce negative staining exclusively of the single or of the double carbon layer type. If the single layer technique is used, the carbon film often folds back on itself forming small regions of double layer images. Likewise even if the double layer technique is used, sizeable regions of single carbon layer images will still be found on the grid in addition to the desired double layer regions.

As is the case with the droplet technique, both negatively and positively contrasted images are seen in single carbon layer-contrasted regions. Figure 2 illustrates the transition between a region of a single carbon layer (at the top) and a double carbon layer (at the bottom) negatively contrasted field of small ribosomal subunits. Similar fields are frequently observed in which the subunits in the single layer region are positively contrasted. Both types of negative stainings are useful for structural studies on ribosomes. The double layer carbon image offers the advantage of providing a very uniform background of thin stain in the regions between subunits and hence makes it easier to observe the antibodies, although this is still possible, but more difficult in single carbon layer regions. Tilting experiments suggest that subunits contrasted in double layer regions are flattened between the carbon layers. The flattening,

---

[1] A 400-mesh grid coated with adhesive is preferred. Adhesive-covered grids can be made by dissolving the adhesive from a 1/8″ strip of Scotch brand transparent tape (not magic mending tape) in 5 ml of chloroform, pouring the solution over grids arranged on the bottom of a petri dish, and allowing the chloroform to evaporate. After drying thoroughly, the grids should be easily removed from the bottom of the dish using finely tipped forceps. If too much adhesive was used and the grids are firmly attached to the bottom of the dish, the dish can be rinsed with chloroform to remove the excess adhesive.

[2] A good absorbent paper is newsprint covered with very dark printing so that the paper fibers are flattened.

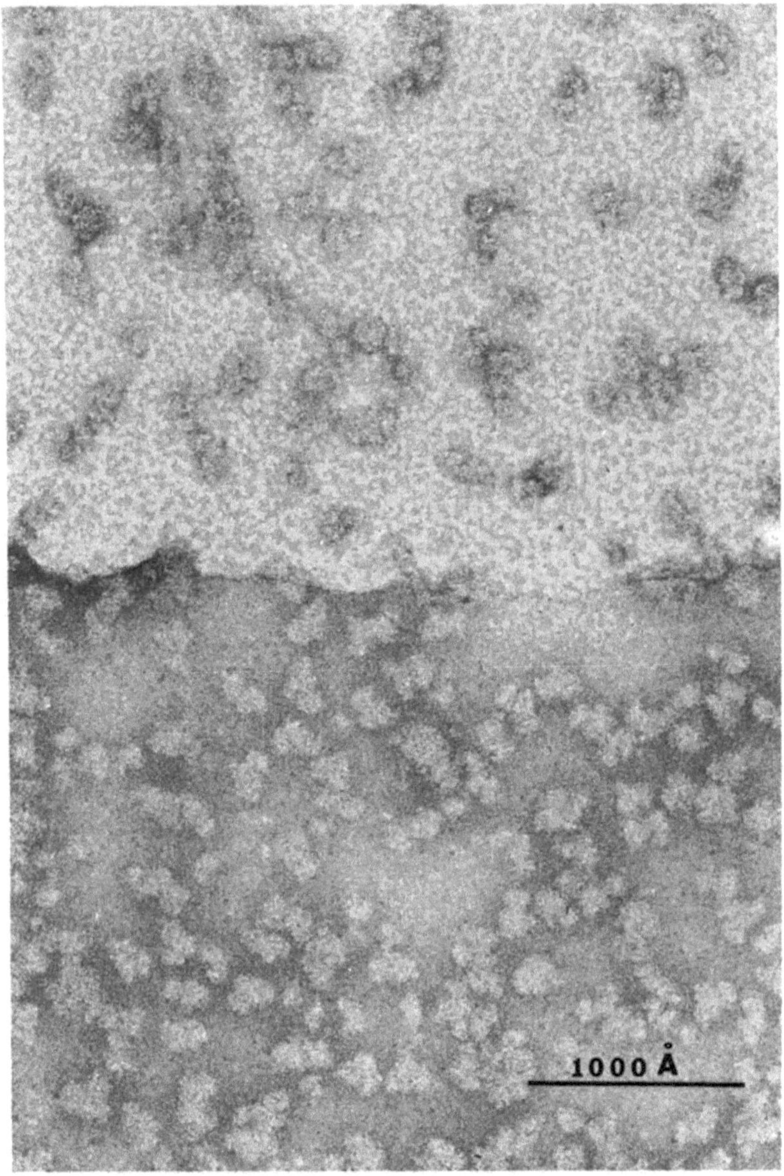

Fig. 2. A negatively contrasted field of *E. coli* small ribosomal subunits. The *lower half of the field* is a double layer carbon image and the upper half is a single carbon layer image

however, does not appear to cause significant distortion of the structure in the plane of the carbon films, since similar structural features of ribosomes are revealed by the droplet, the single carbon layer, and the double carbon layer techniques (LAKE, 1976). The dimensions of ribosomes contrasted by these techniques do vary considerably, however. In Table 1 the lengths of the longest dimension of the small subunit of the *Escherichia coli* ribosome are list-

Table 1. Comparison of ribosomal subunit dimensions determined by different techniques

Technique	Longest Dimension
Electron microscopy (droplet)	180 Å
Electron microscopy (single layer)	220 Å
Electron microscopy (double layer)	250 Å
X-ray scattering from solutions	250 Å

ed for each of the three staining techniques, together with the length measured by X-ray scattering from solutions of small subunits (HILL et al., 1969).

The largest dimension of the 30 S subunit as measured by X-ray scattering involves no assumptions about the shape of the subunit and is a relatively reliable physical measurement of the maximum distance within the subunit in solution. Hence, for ribosomal subunits at least, the double layer carbon method gives images with dimensions that are quite close to those measured in X-ray studies. The single carbon layer method gives dimensions that are about 10% less than the X-ray results, and the droplet method gives dimensions that are about 30% less than the X-ray results.

A major advantage of the double carbon layer technique is that the areas of negative contrast are frequently much larger than those typically obtained with the droplet method. The double layer images can, in favorable circumstances, extend over several squares of a 400-mesh grid with only slight differences in the negative staining occurring throughout the entire area. A typical large area of double layer contrasted 50 S subunits is shown in Figure 3.

Although the double layer technique is obviously the method of choice for antibody-labeling experiments on large and small ribosomal subunits, it is not a panacea. For macromolecular assemblies that are more than 250 – 300 Å thick, the double layer technique tends to flatten the specimens to such an textent that structural details are obscured. For this reason the single layer technique is, in general, superior to the double layer technique for contrasting monomeric ribosomes. In other cases the droplet method is superior to

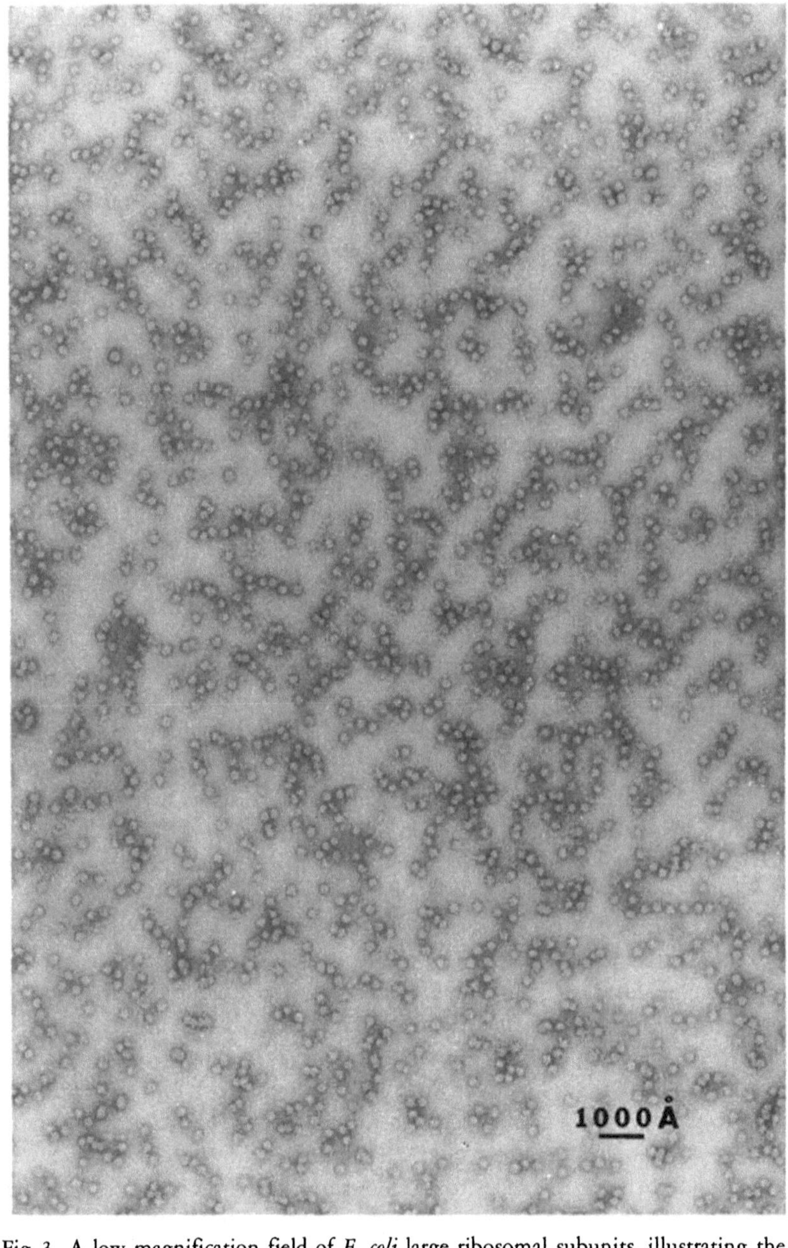

Fig. 3. A low magnification field of *E. coli* large ribosomal subunits, illustrating the large regions of double carbon layer negatively stained images that can be obtained

both the single and double carbon layer methods. For example, in investigating the structure of the capsid of bacteriophage ΦCbK (LEONARD et al., 1972; LAKE and LEONARD, 1974) it was found that droplet negative staining produced electron micrographs in which structural details were preserved to 20 Å resolution as measured by optical diffraction whereas capsids negatively stained by the other two methods did not diffract. The tail of the same phage, however, produced optical diffraction patterns extending to 16 Å and 20 Å resolution when contrasted by the single carbon layer and the droplet techniques, respectively (LEONARD et al., 1973). Hence, the staining properties of a particular specimen cannot be predicted in advance, and in any new situation it is sensible to try all three techniques.

C. Interpretation

Electron microscopy of ribosomes is an extremely useful technique for studying their gross structure. Studies on *E. coli* polysomes, monomeric ribosomes, and subunits by negative contrasting (HUXLEY and ZUBAY, 1960; EISERLING et al., 1964; BRUSKOV and KIESELEV, 1968; LUBIN, 1968; WABL et al., 1973 a, b; WABL, 1974; VASILIEV, 1974; LAKE et al., 1974 a, b, c; LAKE and KAHAN, 1975; LAKE, 1976) and by shadow casting (HALL and SLAYTER, 1959; HART, 1962; 1965; VASILIEV, 1971; 1974) have shown similar ribosomal features that are becoming clearer. In particular, several different developments have recently converged to make the structure of ribosomes better understood. One of these, the application of an objective approach to the interpretation of electron micrographs, has been developing during the last decade and will be emphasized and illustrated in detail in this Chapter. Another important factor has been the increased understanding of the protein chemistry, immunochemistry, and biochemistry of ribosomes. Since structural studies of other supramolecular assemblies are likely to parallel those of the ribosome, a chronologic description of the stages of understanding ribosomal structure is useful to illustrate the way in which our present knowledge has evolved.

In interpreting transmission electron micrographs, it is important to keep in mind that the images differ in a fundamental way from the *reflection* images that are encountered in our everyday visual experiences. Reflection images reveal surface detail from the surface nearest the observer, whereas, in transmission electron microscopy the electrons transverse the entire specimen. In transmission images, structure on the side of the specimen facing the electron beam is contrasted to the same extent as features on the side of the specimen that faces away from the electron source. The contribution of structural features from both the front and back of the structure to the image is a confusing aspect of the *projection* images seen in transmission electron microscopy, and is, perhaps, the most difficult aspect of interpreting electron micrographs.

In negatively contrasted images, contrast is not derived from the molecular complex being imaged, but from the coating of heavy metal that outlines the boundaries of the ribosome and usually does not penetrate very deeply into the structure. In the double carbon layer image, ribosomes are encased in a shell of uranyl acetate that is surrounded on both sides by carbon films. In this symmetric environment, structural details from the near and the far sides of the ribosome are represented in approximately equal amounts in electron micrographs. In specimens contrasted by the droplet method or by the single carbon layer method, ribosomes are in a more asymmetric staining environment since one side of the ribosome faces a carbon substrate and the other faces the air. This can result in different amounts of contraction of the negative stain during drying and can produce unequal contrasting of surface detail from both sides of the image. Obviously these factors should be kept in mind when interpreting images of negatively contrasted molecular complexes, and they will be referred to, implicitly, throughout this Chapter.

The *E. coli* ribosome is in many ways an ideal organelle for structural studies since so much information is available concerning its function and structure through biochemical, immunologic, and genetic studies (for a recent review see *Ribosomes* edited by NOMURA et al., 1974). The smaller *E. coli* ribosomal subunit has a molecular weight of approximately 0.9×10^6 daltons and consists of a single molecule of 16 S RNA and approximately 21 different proteins. The proteins are, for the most part, quite small with molecular weights typically between 10,000 and 30,000 daltons. In capsule form, the principal function of the small subunit is to facilitate and to provide a site for the mRNA-tRNA, codon-anticodon interaction. The larger *E. coli* ribosomal subunit has a molecular weight of approximately 1.55×10^6 daltons and consists of two different molecules of RNA, a 5 S and a 23 S RNA, and approximately 34 different proteins. The proteins, are for the most part, slightly larger than the small subunit proteins. The functions of the large subunit center primarily about peptidyl transfer. Throughout this Chapter, antibody-labeling techniques will be primarily illustrated with reference to the small subunit since considerably more is known about its biochemistry. All of the small subunit proteins have been purified. A number of them have been completely sequenced and all of them have been at least partially sequenced. Antibodies have been prepared against all of the small subunit proteins, and these are specific in the sense that antibodies against one small subunit protein do not cross react against another small subunit protein in Ouchterlony gels (STÖFFLER and WITTMANN, 1971; KAHAN et al., 1974). In addition, the small subunit is a self-assembling structure. Under the proper ionic and temperature conditions, fully functional small subunits can be assembled from proteins and RNA (TRAUB and NOMURA, 1968). There are also considerable functional homologies with respect to small subunit proteins in the prokaryotes. For example, between *E. coli* and *B. stearothermophilus* there is a one-to-one functional correspondence between nearly all small subunit proteins (HIGO et al., 1973). In other words, it is possible to reconstitute a fully functional hybrid small subunit containing 20 of the 21 *E. coli* proteins, *E. coli* RNA, and the remaining protein from *B. stearothermophilus*. Although there is a one to one functional correspondence between these proteins, they are, in general, antigenically different (HIGO et al., 1973). Not only is it possible to reassemble functional subunits, but the sequence of additions of proteins during assembly has been deduced (MIZUSHIMA and NOMURA, 1970) and is summarized in the assembly map shown in Figure 4 (HELD et al., 1974). Arrows between proteins indicate a facilitation of one protein on the binding of a subsequent protein, and a thick arrow indicates a major facilitating effect. For example, the thick arrow from 16 S RNA to S4 indicates that S4, in the absence of other proteins, binds directly to 16 S RNA.

Antibody mapping experiments are quite simple in theory. Since the IgG molecule is bivalent, a single IgG can react with two subunits. Hence subunits are reacted with antibodies against a particular protein, pairs of subunits connected by IgGs are separated, and then are negatively stained by the double carbon layer technique for electron microscopy. Images of subunits con-

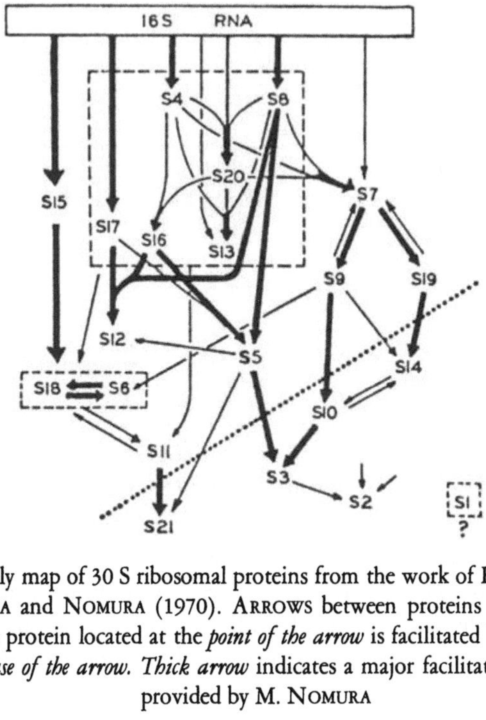

Fig. 4. Assembly map of 30 S ribosomal proteins from the work of HELD et al. (1974) and MIZUSHIMA and NOMURA (1970). ARROWS between proteins indicate that the binding of the protein located at the *point of the arrow* is facilitated by the protein located at the *base of the arrow*. *Thick arrow* indicates a major facilitating effect. Figure provided by M. NOMURA

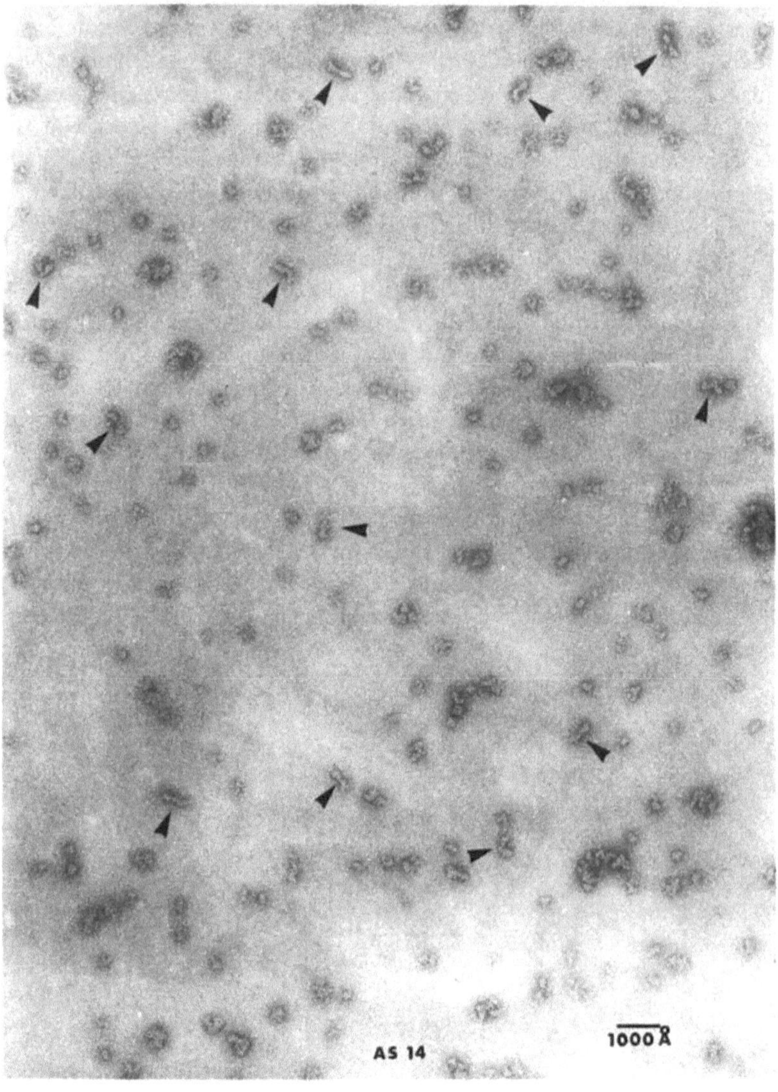

Fig. 5. A low magnification, double carbon layer image of a field of 30 S subunits reacted with AS14 antibodies. From Lake et al. (1974 b)

nected by an IgG molecule are easily recognized, and the presence of two subunits connected by the same antibody provides a convenient redundancy to the data. Images showing the site of antibody attachment to subunits seen in different orientations permit the region of antibody binding to be mapped in three dimensions.

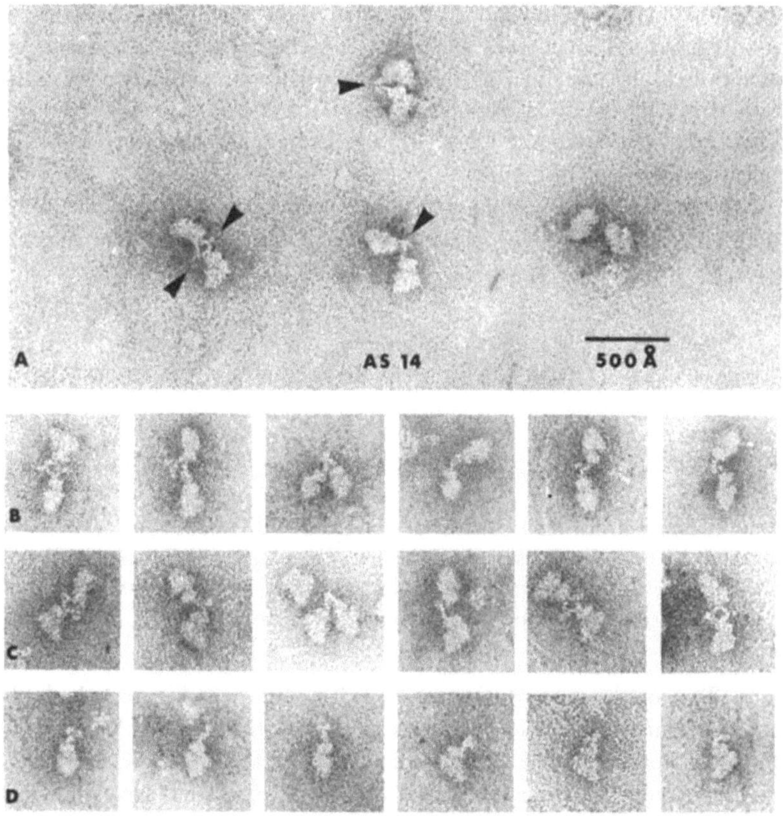

Fig. 6 A – D. Electron micrographs of small subunits reacted with AS14 antibodies. The antibodies attach only to a single region in the upper one-third of the subunit. The Fab regions and the Fc region of the antibodies are both visible. (A) A field showing pairs of subunits linked by one or two antibody molecules (*arrows*). (B) A gallery of dimers in which the quasi-symmetric profile is clearly identifiable (the lower subunit in each case). In the two frames at the left two antibody molecules are visible, bound to both subunits or to one subunit. (C) A gallery of dimers in which the asymmetric profile is visible (the lower subunit in each case). Two bound antibodies are visible in the frame at the left and in the two frames at the right. (D) A gallery of monomers having a single bound AS14 antibody. The three frames at the left show monomers in the quasi-symmetric profile; the three frames at the right show monomers in the asymmetric profile. From LAKE et al. (1974 b)

The first small subunit protein to be mapped by antibody binding was S14 (LAKE et al., 1974 b). S14 is a functional protein that is assembled into the small subunit late in the assembly process. If it is omitted from the reconstitution the resulting subunits are functionally inactive (NOMURA et al., 1969). A lower power field of small subunits reacted with AS14 (IgG against protein S14) is shown in Figure 5. Inspection of the micrograph reveals many pairs of subunits connected by antibodies. A higher mangnification region of another field is shown in Figure 6 A. In this field one can see pairs of subunits connected by either one or by two IgGs (indicated by arrows). In the central pair of subunits the Fab region attached to each subunit is visible as well as the Fc region of the IgG.

The small subunit profiles observed in electron micrographs correspond to the small subunits being positioned on the carbon in different orienta-

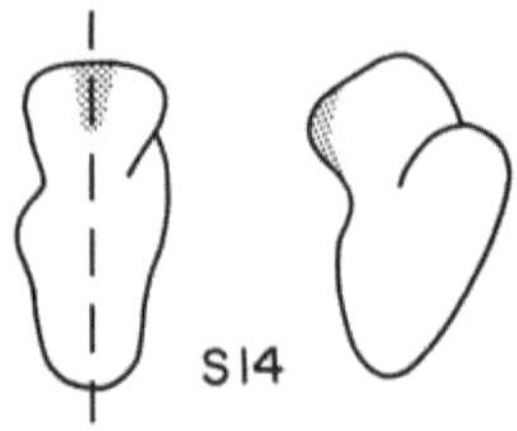

Fig. 7. A diagrammatic representation of the region of attachment of antibodies to protein S14. The site is represented by stippling in the quasisymmetric view (*left*) and in the asymmetric view (*right*). The line of approximate mirror symmetry in the quasisymmetric view is represented by a dashed line

tions. Two of the most distinctive views are shown in Figure 7. In both views the subunit is divided into an upper 'one-third' and a lower 'two-thirds' by a partition. The view illustrated at the left of Figure 7, the 'quasisymmetric' view, is characterized by a line of approximate mirror symmetry that is coincident with the long axis of the subunit and is indicated by a dashed line. Pairs of subunits are shown in Figure 6 in which the lower subunit of the pair is in this orientation. As illustrated at the left of Figure 7, AS14 antibodies attached at the top of the upper one-third on the line of approximate mirror symmetry. The lower subunit of each pair in Figure 6 B is oriented in the 'asymmetric' view (the view shown on the right in Fig. 7). The location of S14 in the upper one-third, on the concave surface of the subunit, is illustrated on the right side of Figure 7.

Antibodies against S14 thus mapped at a single, unique three-dimensional site. Although it had been shown that these antibodies would react with purified *E. coli* protein S14 but with no other purified *E. coli* small subunit protein, the possibility existed that the conformations of ribosomal proteins could be altered when they were incorporated into the small subunit, and in this altered conformation they might react with AS14 antibodies. To demonstrate that this was not happening and that antibody binding was, in fact, dependent on the presence of *E. coli* S14 in the ribosome, we replaced E. S14 (S14 from *E. coli*) with its functional counterpart from *B. stearothermophilus,* B. S14, in hybrid reconstituted subunits. Hybrid subunits containing all of the *E. coli* RNA and proteins except S14, which was replaced by B. S14, were reconstituted, as were subunits that contained all of the *E. coli* RNA and proteins. Although E. S14 and B. S14 are functionally equivalent, they are antigenically distinct. When these subunits were reacted with AS14 antibodies (see Fig. 8), as expected, AS14 antibody binding to the hybrid B. S14 subunits was only 9% of that observed with the *E. coli*-reconstituted subunits. It was concluded that antibody binding to the subunits was dependent upon *E. coli* S14 being present in the subunit (LAKE et al., 1974 c).

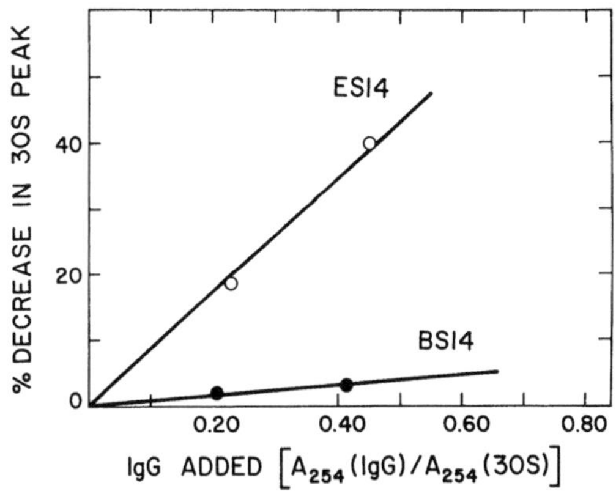

Fig. 8. *E. coli* S14-dependent aggregation of reconstituted subunits by AS14 antibody. Subunits were reacted with AS14 and sedimented through sucrose gradients, and the amount of monomer 30 S ribosomes was determined as the A_{254} of the monomer peak. The AS14-dependent aggregation of ribosomes containing *E. coli* S14 (○) and the corresponding *B. stearothermophilus* protein (●) are shown expressed as percentage decrease from the monomer peak height obtained in the absence of IgG

In retrospect, there are other, more compelling, reasons for performing the hybrid reconstitution controls (LAKE and KAHAN, 1975). One of the most serious is that the Ouchterlony precipitin test requires that antibodies react with a minimum of three different antigenic determinants on a protein in order for a visible precipitating lattice to be detected. In other words, it is possible, in theory, that an antibody against ribosomal protein X, could react with two determinants of each of the remaining ribosomal proteins and yet none of these cross reactions would be detected on Ouchterlony gels. Hence, lack of cross reactivity on Ouchterlony gels is not proof that antibodies do not react with more than one protein. A number of homologies, four residues long, have been reported (see e.g., HIGO and LOERTSCHER, 1974; WITTMANN and WITTMANN-LIEBOLD, 1974) between the sequences of different *E. coli* ribosomal proteins. This is sufficiently long to form an antigenic determinant (see e.g., KABAT, 1968) so that if these regions of the proteins are lo-

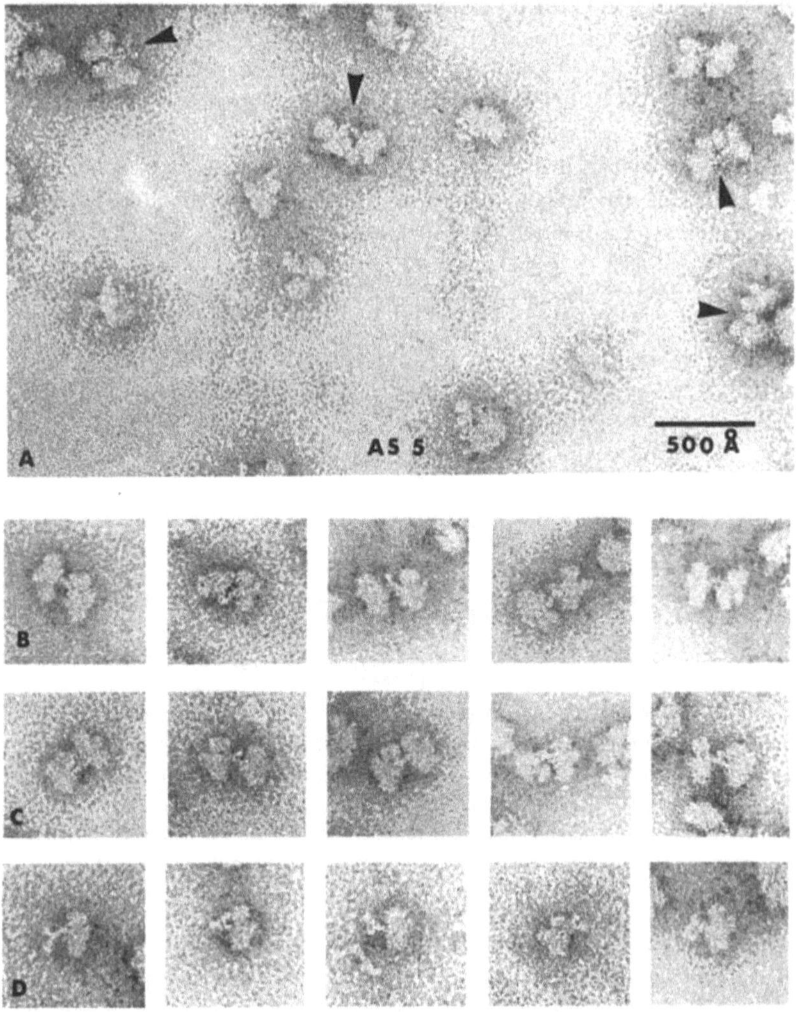

Fig. 9. A – D. Electron micrographs of small subunits reacted with AS5 antibodies. The antibodies attach to a single region of the subunit at, or just below, the partition. (A) A field showing pairs of subunits linked by single antibody molecules (indicated by *arrows*). (B) A gallery of pairs in which the quasi-symmetric profile is clearly identifiable. (C) A gallery of pairs in which the asymmetric view is identifiable. (D) A gallery of monomers having a single bound AS5 antibody. From LAKE and KAHAN (1975)

cated at the surface of the subunits some cross reactions are likely to occur. Hence hybrid reconstitution experiments are needed to demonstrate that the antibody binding seen in the electron-microscopic studies is dependent upon the presence of the corresponding *E. coli* ribosomal protein within the subunit. In structural studies of macromolecular assemblies, other than ribosomes, detection of cross reaction can be a problem since reconstitution conditions have not, for the most part, been worked out. In those instances, the possibility of cross reactions that are undetected in Ouchterlony gels should be kept in mind, particularly for antibody binding sites that map at multiple locations.

The second protein to be described, S5, is also a functional protein. The localization of this protein is interesting because it maps at a different region of the small subunit than the other proteins that will be discussed in this Chapter and because it is also the only protein, of those to be discussed here, that has not been observed to be cross-linked to initiation factors. AS5 antibodies attach at a single site as shown in Figure 9 (LAKE and KAHAN, 1975). In the field in Figure 9 A, numerous pairs of subunits linked by antibodies are indicated by the arrows. The subunits are shown in the quasi-symmetric view in Figure 9 B and monomers in this view are shown in the first two frames of 9 D. In the quasi-symmetric view, antibodies attach to the lower two-thirds just below the partition (and on the opposite side of the quasi-mirror line from the high shoulder that will be discussed in detail in reference to protein S13). In the asymmetric view (Fig. 9 C and last two frames of 9 D) the antibodies bind on the concave surface of the subunit just below the partition as shown in Figure 10. As with S14, substitution of B. S5 for E. S5 reduced antibody binding and demonstrated that AS5 binding is dependent upon *E. coli* S5 being present in the subunit.

At this stage, we began to investigate other types of ribosomal proteins to see if they might have different structural properties that were reflected in their distributions on the small subunit surface. Protein S4 was the first to be investigated since it is a typical assembly protein. It binds directly to 16 S RNA (see the assembly map in Fig. 4) and thereby facilitates the subsequent incorporation of at least four other proteins into the subunit. If S4 is omitted, then the subunits that are formed sediment at 20 to 25S (NOMURA et al., 1969) (instead of the normal, 30 S), indicating that the assembly has

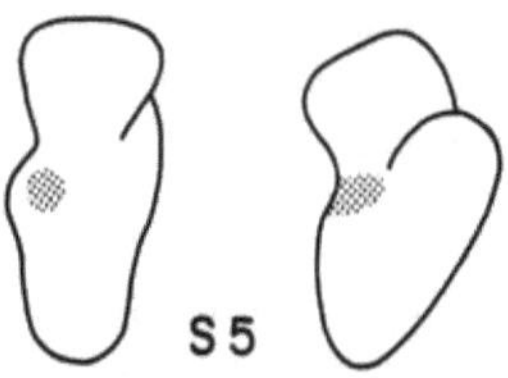

Fig. 10. A diagrammatic representation of the site of attachment of antibodies to protein S5. The site is represented by stippling in the quasi-symmetric view (*left*) and in the asymmetric view (*right*)

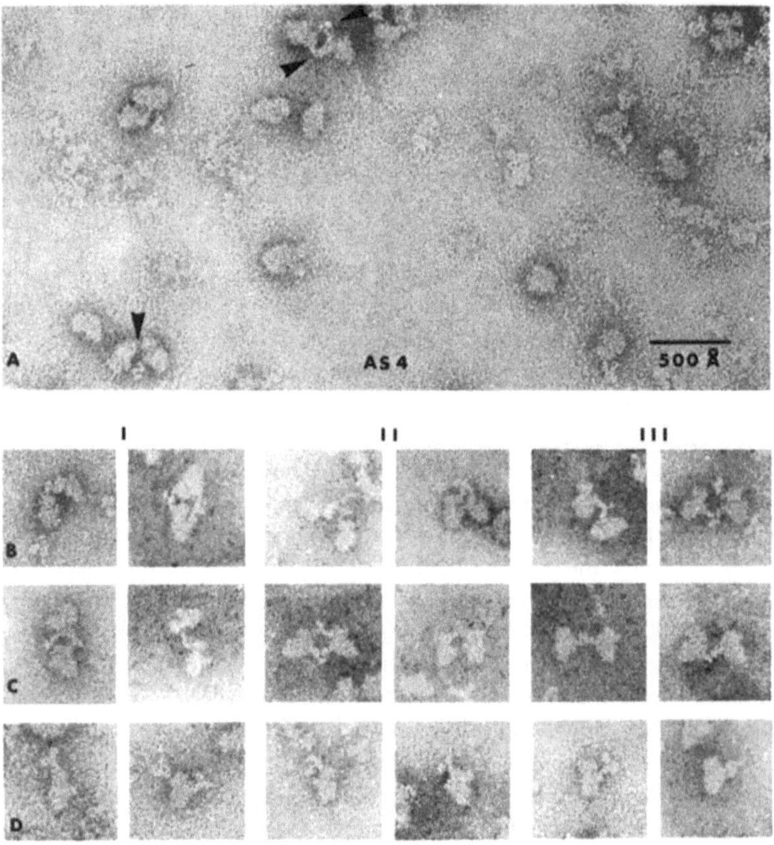

Fig. 11. A – D. Electron micrographs of small subunits with AS4 antibodies. The antibodies attach to three regions of the subunit, shown schematically in Figure 12. Sites *I, II,* and *III* are shown in the first, second, and third paired columns, respectively. Simultaneous bindings of antibodies to sites *I* and *III, I* and *II,* and *II* and *III* are shown in *IB* frame *2, IC* frame *1,* and *IIC* frame *1,* respectively. (A) A field showing pairs of subunits with antibodies attached (*arrows*) at Sites *III, I,* and *II* from left to right. (B) Dimers in which the quasi-symmetric profile is clearly shown (the lower subunit(s), in each case). (C) Dimers in which the asymmetric profile is visible (the lower subunit in each case). (D) Monomers having single bound AS4 antibodies. The first frame of each site shows the quasi-symmetric profile and the second frame shows an asymmetric profile. From LAKE et al. (1974 c)

not proceeded to form a subunit having the correct quaternary structure. When AS4 antibodies were reacted with subunits, we were surprised to find that they did not bind at a single site on the subunit surface but instead reacted at three different regions of the subunit (LAKE et al., 1974 b, c) (see Fig. 11 A). Antibodies linking pairs of subunits attached to these three sites (labeled I, II, and III) are shown in the quasi-symmetric view in 11 B and in the asymmetric view in 11 C, and antibodies attached at these sites to single subunits are shown in 11 D. Sites I and II are in the upper one-third and site III is located in the lower two-thirds, below the partition. In the quasi-symmetric view, sites II and III map on the same side of the subunit with respect to the approximate mirror line and on the side opposite site I (see Fig. 11 B, second frame from the left). Pairs of subunits connected by two antibodies that were attached to different sites were particularly useful. All of the possible combinations of pairs of sites were found to be stearicly possible. Pairs of subunits with antibodies simultaneously attached to sites I and II (Fig. 11 C, frame 1), I and III (Fig. 11 B, frame 2), and sites II and III (Fig. 11 C, frame 3) are illustrated. The locations of the S4 sites are shown diagrammatically in Figure 12.

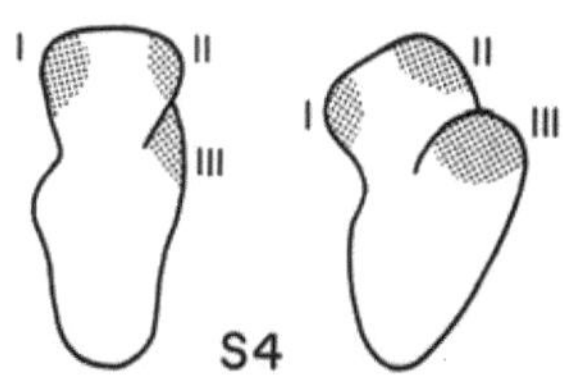

Fig. 12. A diagrammatic representation of the three regions of attachment of antibodies to protein S4. The regions are indicated by stippling in the quasi-symmetric view (*left*) and in the asymmetric view of the subunit (*right*)

For subunits linked by antibodies attached to two sites, it was important to note that an antibody that attached to a given site on one subunit always attached to the same site on the other subunit. No crosses were seen between sites indicating that the sites are antigenically distinct. Protein S4 is known to be present in one copy per subunit. Since the binding at each of the three sites was dependent on the presence of *E. coli* S4 (LAKE et al., 1974 C), and since each of the S4 sites are antigenically distinct, it was concluded that S4 probably exists *in situ* in an elongated conformation. In order for S4 to extend to all three regions it must be at least 170 Å long. S4 has been completely sequenced (REINBOLT and SCHLITZ, 1973): it consists of 203 residues, and if the protein were entirely α-helix, would extend approximately 305 Å. There is, of course, no reason to require that S4 be α-helix; the calculation simply suggests that it is sufficiently large to account for its observed distribution in the subunit. Mapping the amino and carboxyl terminal regions of S4, employing antibodies against fragments of S4, remains to be completed. Hence the conclusion that S4 is elongated is still tentative.

One question that the assembly map of NOMURA and his Co-workers raises is the relationship that the interactions shown on the map have to the spatial relationships observed between proteins. For example, what are the

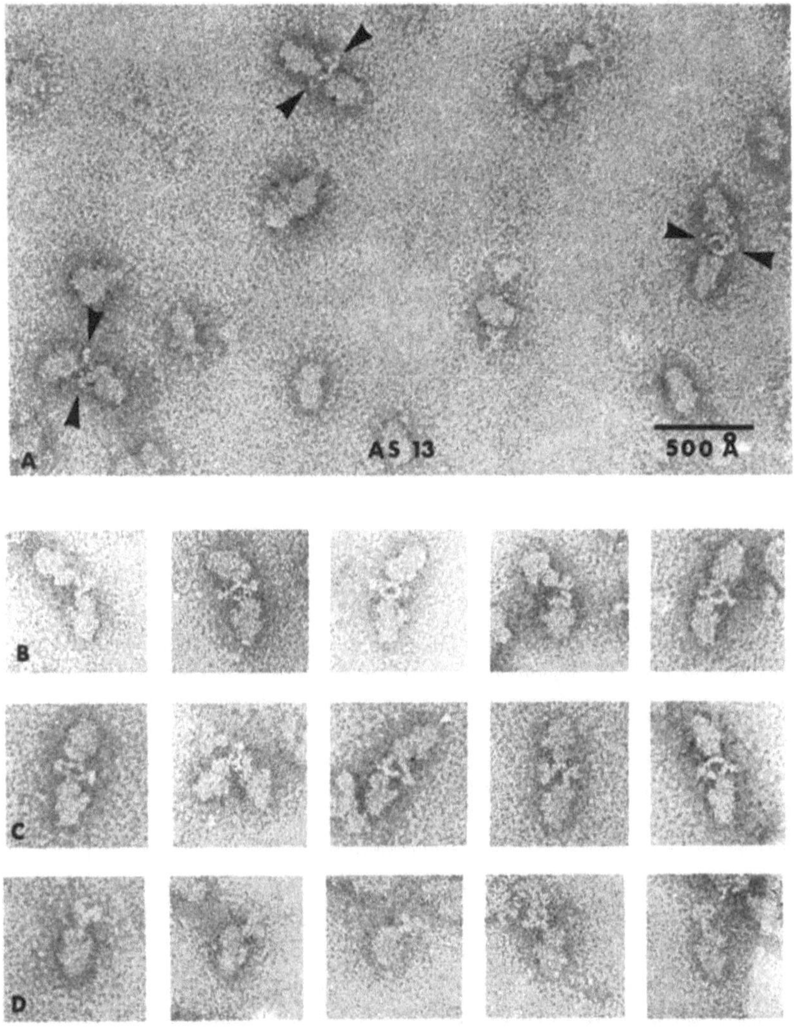

Fig. 13. A – D Electron micrographs of small subunits reacted with AS13 antibodies. The antibodies attach to a single region of the subunit. (A) A field showing pairs of subunits linked by two antibody molecules (*arrows*). (B) A gallery of pairs of subunits in which the quasi-symmetric profile is clearly identifiable (*lower subunit* in each case). (C) A gallery of pairs of subunits in which the asymmetric profiles are visible. (D) A gallery of monomers having a single bound AS13 antibody. The three frames at the left show monomers in the quasi-symmetric profile and the two at the right show the asymmetric profile. From LAKE and KAHAN (1975)

physical interactions that occur when the binding of a protein facilitates the subsequent incorporation of another protein? Are both proteins adjacent or perhaps does the binding of one protein cause a movement of a part of an RNA chain or another protein that may travel down the chain and allow the subsequent addition of the second protein at a region perhaps 100 Å, or so, from the site of the first protein? Obviously the answer to this will require

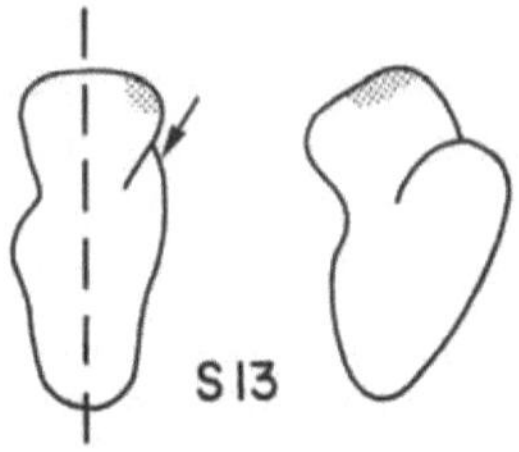

Fig. 14. A diagrammatic representation of the site of attachment of AS13 antibodies. The site is represented by stippling in the quasi-symmetric view (*left*) and in the asymmetric view (*right*). In the quasi-symmetric view the line of approximate mirror symmetry is dashed, and the high shoulder is indicated by an arrow

the detailed mapping of all the proteins in the small subunit. Preliminary mapping studies on the locations of all the small subunit proteins and complete mappings (including hybrid reconstitution controls) on approximately one-half of the small subunit proteins suggest that an interaction on the assembly map usually corresponds to both proteins being located in the same region of the subunit. The converse statement, that physically adjacent proteins are related by assembly map interactions is, of course, not true. Consider, for example, the relationships between S4, S13, S14, and S19 that will be described below.

Addition of protein S13 to the nascent subunit is facilitated by the prior incorporation of protein S4 and by the incorporation of several other proteins (see Fig. 6). S13 maps at a single region of the subunit, as shown in the galleries in Figure 13 B – D. In the quasi-symmetric view, Figure 13 B, S13 maps in the upper one-third well to the side of the approximate mirror line as shown in the left panel of Figure 14. In the asymmetric view, Figure 13 C and the right panel of Figure 14, S13 maps near the convex surface and in the upper one-third. The mirror line seen in the images is only approximate and a high shoulder, shown by the arrow in Figure 14, breaks the symmetry and serves as a convenient feature to distinguish between the sides of the quasi-symmetric view. The high shoulder is approximately 30 Å higher than the opposite shoulder, and an accumulation of stain often separates it from the upper one-third portion of the subunit at the level of the partition. In the quasi-symmetric view, the S13 site maps on the same side as the high shoulder.

To map proteins uniquely in three dimensions with respect to the 30 S structure, it was necessary to identify reproducibly the hand of the small subunit in the quasi-symmetric projection. To do this it was necessary to provide convincing evidence that the major symmetry-breaking feature of this view (the high shoulder) was constant and reproducible. Protein S13 was chosen as a fixed reference marker for these purposes since in the quasi-symmetric view it maps well off the mirror line in the one-third region and above the high shoulder. To illustrate the reproducibility of the high shoulder in the quasi-symmetric view, 100 subunits (linked to another subunit by one or two antibodies) were selected in which the subunit and Fab segment of the linking antibody were clearly defined, the subunit was oriented in the quasi-symmetric projection, and a high shoulder could be discerned. These 100 subunits were then scored according to whether the high shoulder was on the left or the right, and then scored a second time as to whether the antibodies were attached on the left or the right side of the approximate mirror line. The results are shown in Table 2.

Table 2. Distribution of AS13 antibodies with respect to the high shoulder of the 30 S subunit in the quasi-symmetric view

Position of antibody	Number of images
Antibody on same side as high shoulder	94
Antibody on side opposite high shoulder	3
Antibody not clearly on either side of subunit	3
	100

In 94 of the 100 images, the shoulder and AS13 antibodies were located on the same side of the subunit. In three images the site of antibody attachment could not be clearly assigned to one side of the subunit. In the remaining three images (see e.g., Fig. 15 C and F, last frames), the shoulder was incorrectly identified. The results demonstrate that the high shoulder is a reproducible structure that maps in a consistent manner with respect to protein S13.

Thirty of these images are shown in Figure 15. Rows A, B, and C show antibodies attached to the left of the approximate mirror line and rows D, E, and F illustrate antibodies attached to the right of the approximate mirror line. The high shoulder coincides with the side to which the AS13 antibodies attach; however, two exceptions have been included and are illustrated in the right frame of C and the right frame of F. In pairs of subunits joined by two

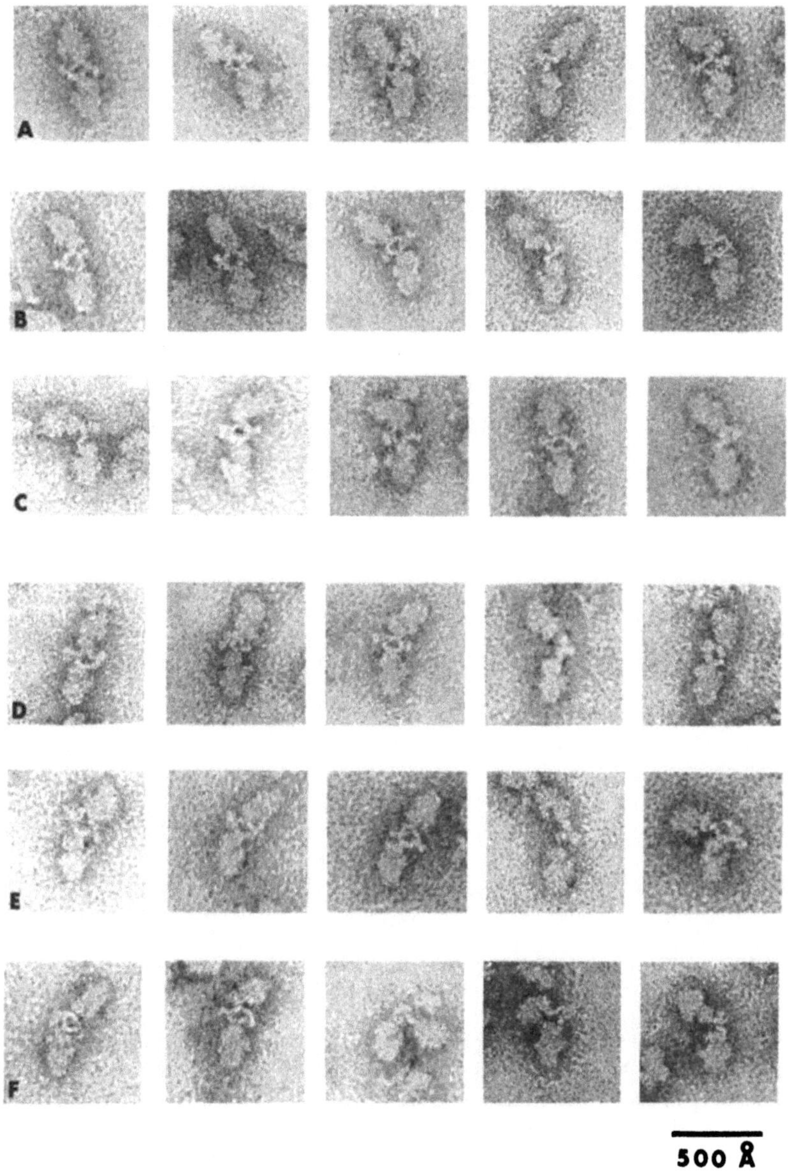

Fig. 15. A – F. A gallery of pairs of small subunits with the lower subunit oriented in the quasi-symmetric projection and linked by AS13 antibodies. Rows (A), (B), and (C) illustrate antibodies attached to the left of the approximate mirror line and rows (D), (E), and (F) illustrate antibodies attached to the right. The high shoulder is on the same side of the approximate mirror line as the antibodies. Two exceptions have been included. These are shown in the last (*right*) frames in rows (C) and (F). From LAKE and KAHAN (1975)

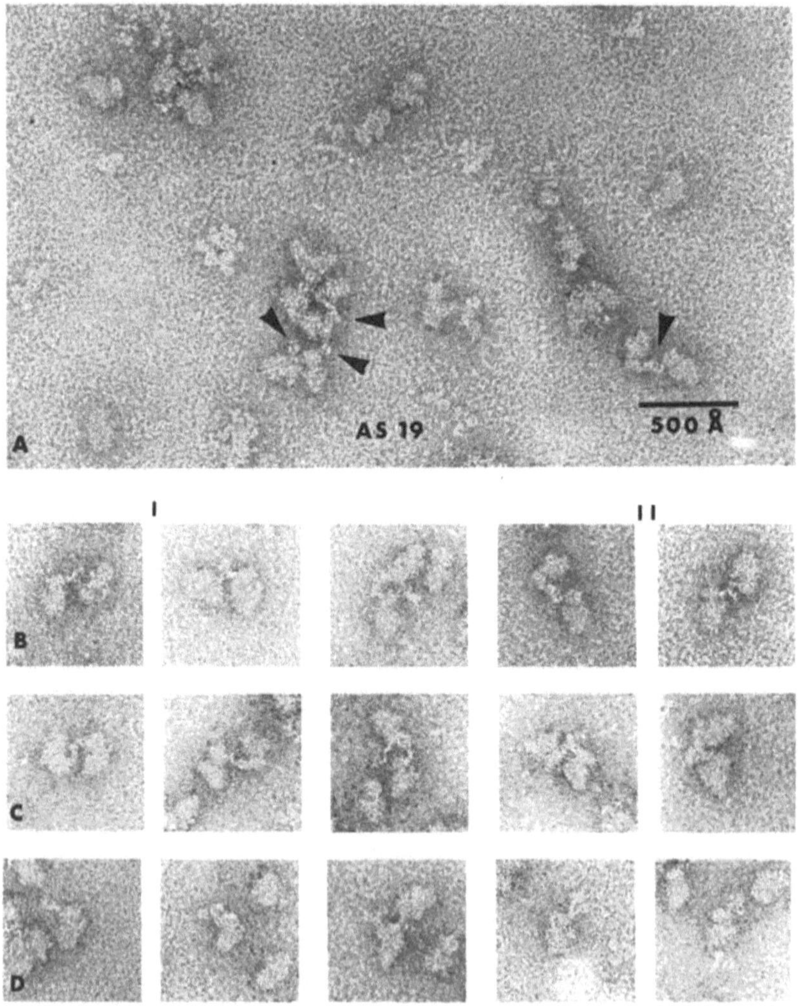

Fig. 16. A – D. Electron micrographs of reconstituted subunits reacted with AS19 absorbed with S13. Note that all subunits in the Plate are hybrid reconstituted subunits containing *E. coli* 16 S RNA and proteins, except for E. S13, which is replaced with its *B. stearothermophilus* functional equivalent, B. S13. The antibodies attach to two regions of the subunits shown schematically in Figure 17. Sites *I* and *II* are shown in the two left and the two right columns, respectively. Simultaneous binding of antibodies to both sites is illustrated in the center column. (A) A field of subunits linked by AS19 antibody (indicated by *arrows*). An aggregate consisting of several subunits is indicated by the group of 3 arrows near the center of the Plate. (B) Pairs of subunits in which the quasi-symmetric profile is clearly shown (the *lower subunit* in all cases). (C) Pairs of subunits in which the asymmetric profile is clearly shown. (D) Monomers having single bound AS19 antibodies, except for the third frame from the left, which shows a pair of subunits linked by AS19(*II*) antibody with a single AS19(*I*) antibody attached to the lower subunit of the pair. The first and fourth frames from the left show quasi-symmetric profiles; the second, third, and fifth frames from the left show asymmetric profiles. From LAKE and KAHAN (1975)

antibodies, the same orientation of the entire complex is consistently seen, indicating that the pairs are held together quite rigidly. These images are consistent with a dyad axis oriented approximately parallel to the plane of the carbon substrate and coinciding with the dyads of both IgG molecules.

Protein S13 mapped adjacent to S4 region II in a manner consistent with the assembly map interaction (see Fig. 6) and with cross-linking results (SOMMER and TRAUT, 1974), but at a distance of approximately 65 Å from S14. At the time, this was difficult to understand since S14 and S13 had both been shown to be neighbors of S19 in cross-linking studies using bifunctional reagents (LUTTER and KURLAND, 1973; SUN et al., 1974; and BODE et al., 1974), and it had been thought that therefore S14 and S13 were also neighbors.

The antibody-labeling experiments had pointed out a potential inconsistency that suggested that there might be something unusual about S19, and made S19 a particularly interesting protein to map. It was necessary to use reconstituted subunits for the S19 mapping experiments for technical reasons, so that all of the subunits shown in Figure 16 are reconstituted. In addition, they are also *hybrid* reconstituted subunits and contain S13 from *B. stearothermophilus.* Nevertheless, the observed structures are indistinguishable from those of derived *E. coli* subunits. Antibodies against S19 reacted to form pairs of subunits, and also formed a relatively high percentage of linear aggregates. Inspection of the linear aggregates (see the arrowed aggregate in the field in Fig. 16 A) revealed antibodies attached to both sides of the upper one-third in the quasi-symmetric view. Galleries of these two sites are shown in the quasi-symmetric view (Fig. 16 B) and in the asymmetric view (Fig. 16 C). Antibodies simultaneously attached to both sites are shown in the central frames of rows B, C, and D, again demonstrating that both sites are antigenically different determinants and that protein S19 is extended. These two AS19 sites, separated by approximately 70 Å, and their locations are illustrated in Figure 17. Site I of S19 maps near S14, and S19-II maps near S13, so that the S13 – S19 cross link is most likely between S13 and S19-II and the S14 – S19 cross link between S14 and S19-I. This is also consistent with the fact that no S13 – S14 cross link has yet been reported.

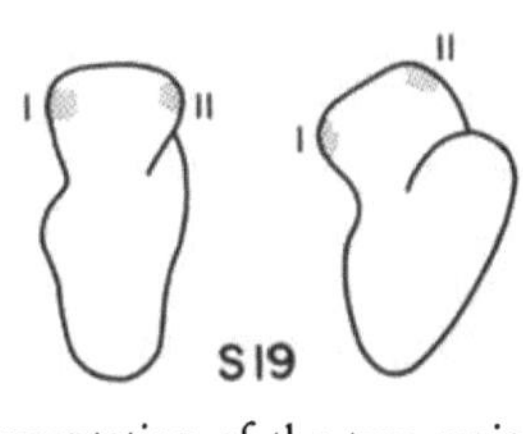

Fig. 17. A diagrammatic representation of the two regions of attachment of antibodies to protein S19. The regions are indicated by stippling in the quasi-symmetric view (*left*) and the asymmetric view (*right*)

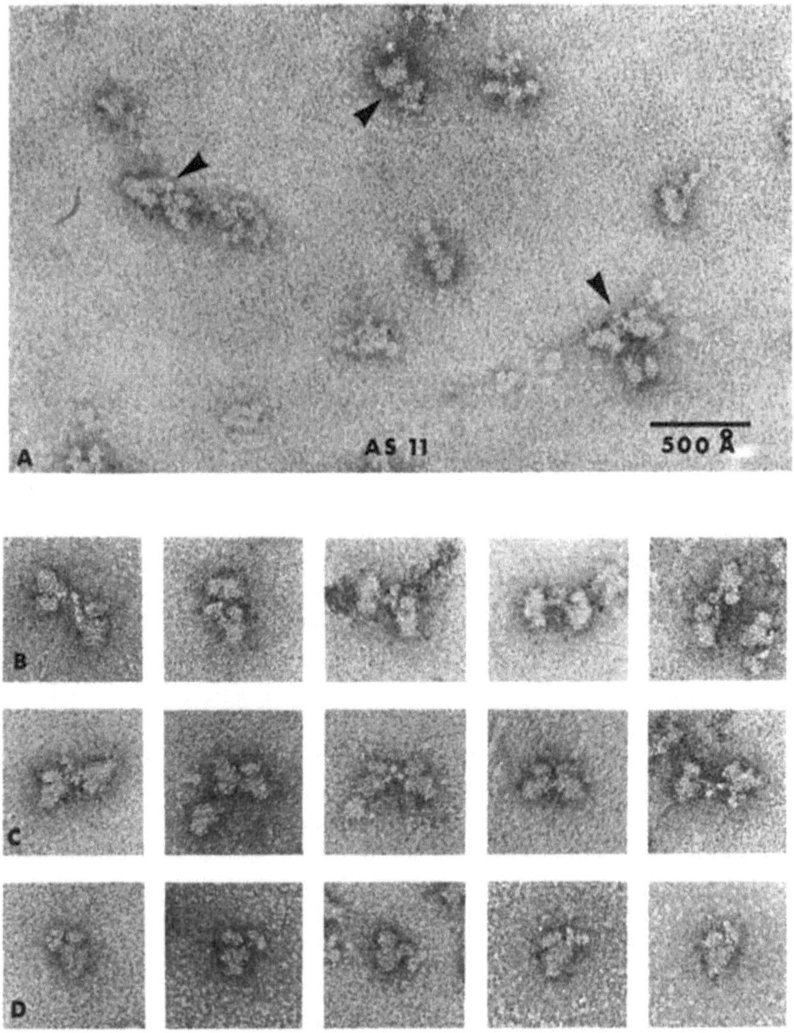

Fig. 18. A – D. Electron micrographs of small subunits reacted with AS11 antibodies. The antibodies attach to a single region of the subunit. (A) A field showing pairs of subunits linked by antibodies (indicated by *arrows*). (B) A gallery of pairs in which the quasi-symmetric profile is clearly identifiable. (C) A gallery of pairs in which the asymmetric profile is visible. In the second frame from the left antibodies are obscured by the subunits and in the last frame a second antibody is attached to one of the subunits. (D) Monomers having a single bound antibody. The three monomers on the left show the quasi-symmetric profile and the two monomers on the right are oriented in the asymmetric profile. From LAKE and KAHAN (1975)

Not only has there been a close correlation between cross-linking experiments and antibody-labeling studies, but cross-linking of initiation factors has also correlated well with the antibody studies and is starting to increase our understanding of the mechanisms of ribosomal function. Initiation factor IF3 may be efficiently cross-linked to S11, S12, S13, and S19 (HEIMARK, KAHAN, JOHNSTON, HERSHEY, TRAUT: manuscript in preparation). IF3 is required for binding a natural messenger to the small subunit, and by localizing it on the surface of the subunit, we hoped to learn something about the site of mRNA binding. Proteins S11, S12, S13, and S19 have all been localized.

Since S11 maps in a different site than S13 and S19 and since it serves to illustrate the structure of the platform region of the small subunit, its location will be described in detail. A field of subunits reacted with AS11 is shown in Figure 18 A. The antibodies react at the single region of the subunit on the platform as shown in Figure 19. In order to describe the location of the antibody binding site it was useful to employ another view of the subunit, the +50° view that is shown in the middle frame of Figure 19. This

Fig. 19. A diagrammatic representation of the region of attachment of antibodies to S11. The site is represented by stippling in the quasi-symmetric (*left*), intermediate, +50° (*middle*), and asymmetric (*right*) projections

view is intermediate between the quasi-symmetric, 0° view, and the asymmetric, +110° view, and in this view the platform is well resolved from the upper one-third of the subunit. Although it was possible to show that S11 mapped on the platform, it was not possible to decide if S11 was on the side of the platform that faced the upper one third or if it were on the side facing away. Since the platform is not much thicker than 30 to 40 Å, it is likely that S11 is exposed on both sides of the platform. Thus the proteins S11, S13, S19(II), and S12 (KAHAN and LAKE, unpublished data) that can be efficiently cross-linked to IF3 all map in the same vicinity on the surface of the small subunit. The region that is implicated in IF3 binding (Fig. 20) includes part of the upper one-third, the platform, and the cleft formed between the upper one-third and the platform.

Subsequently the locations of all 21 small subunit proteins shown in Figure 20 have been mapped in preliminary studies (LAKE and KAHAN, manuscript in preparation) and complete reconstitution controls have been performed for approximately one-half of the 30S proteins. These locations are quite preliminary and for many of the proteins it is possible that new sites will be detected and other sites will disappear as the hybrid reconstitution controls are completed for each of them. Also only very general locations are shown to indicate the regions of the 30S surface where proteins are mapped. For example, if more space were available protein S13 should have been placed near S7-I in the illustration at the right of Figure 20.

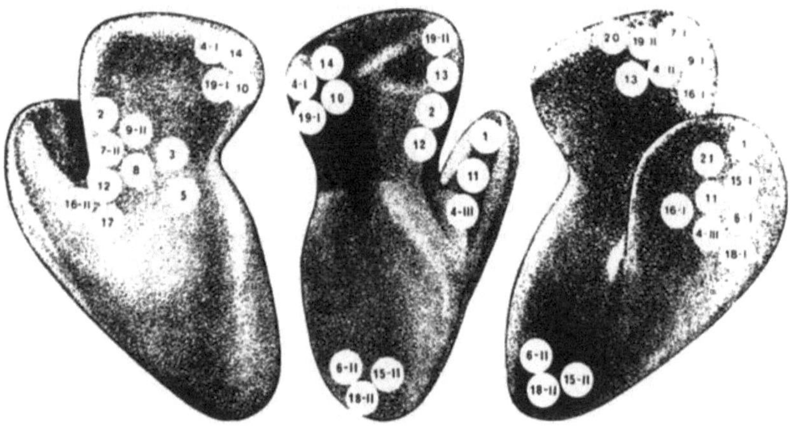

Fig. 20. Preliminary locations of all 21 small subunit proteins. (From LAKE and KAHAN, manuscript in preparation)

Identification of the locations of these proteins has also allowed us to indirectly locate the binding sites of other ligands (LAKE and KAHAN, 1975). Initiation factor IF2 can be cross-linked (in the presence of IF1 and IF3) to S1, S2, S11, S13, S14, and S19 (BOLLEN et al., 1975). The experiments did not distinguish between a direct cross link between IF2 and each of the proteins or between an indirect cross-linking involving a protein bridge of intervening protein or initiation factor. Since a S14-S19 cross link has been reported, S14 might not be directly linked to IF2. With the exception of S14, however, all the remaining proteins map in the same general area as the region of IF3 binding. Since binding of the initiation tRNA to small subunits is dependent on the presence of IF2, this suggests that the region shown in Figure 21 is also the site of initiator tRNA binding.

The locations of the proteins described above also suggest that the codon-anticodon interaction occurs in the cleft between the platform and the upper one-third. When single component omission experiments (NOMURA et al., 1969) were performed, and the loss of various ribosomal functions was correlated with the missing component, it was found that most of the ribosomal functions that could be assayed were cooperative. The absence of a single component interfered with many different functions. Proteins S11 and S12 were two exceptions to this rule. When S11 was omitted, under conditions where messenger misreading was increased, then misreading increased even more. Under the same conditions, when S12 was omitted, misreading was decreased. Thus S11 and S12 are thought to occur near the codon-anticodon interaction site. S11 and S12 are both located in the vicinity of the cleft sug-

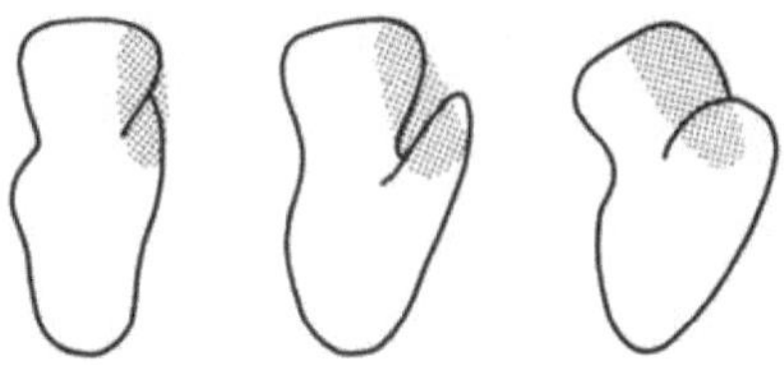

Fig. 21. A diagrammatic representation of the region of the subunit containing proteins that can be cross-linked to initiation factor IF3. The region is indicated by stippling in the quasi-symmetric view (*left*), an intermediate, +50° view (*middle*), and the asymmetric view (*right*)

gesting that this is the region of the subunit where the anticodon of the tRNA is located.

In collaboration with W. STRYKHARZ and M. NOMURA at the University of Wisconsin we have been mapping the locations of large subunit proteins. The approach is similar to that used for the small subunit. If two characteristic profiles corresponding to approximately orthogonal orientations of the subunit can be identified, then the attachment sites of antibodies in these two subunit orientations allow the protein to be mapped in three dimensions. A field of 50S subunits reacted with anti-L7/L12 antibodies is shown in Figure 22 a. The left subunit of each pair in Figure 22 b is in an orientation corresponding to that shown at the left of Figure 23. Likewise the left subunit of each pair in Figure 22 c is in an orientation corresponding to that shown at the right of Figure 23 (i.e. in a projection approximately orthogonal to that in Figure 22 b). Hence proteins L7/L12 were localized on the elongated rod-like appendage of the large subunit as shown in Figure 23. The two proteins L7/L12 have properties that make them unique among ribosomal proteins. They are both identical except that L12 has an acetyl group on its N-terminal serine and L7 does not. L7 and L12 are both required for GTP hydrolysis in the EF-Tu and the EF-G dependent reactions of the elongation microcycle of protein synthesis. Hence their localizations at a single region of the large subunit is likely to have important consequences for understanding the molecular functioning of the ribosome. We have presently mapped the locations of approximately 40% of the large subunit proteins.

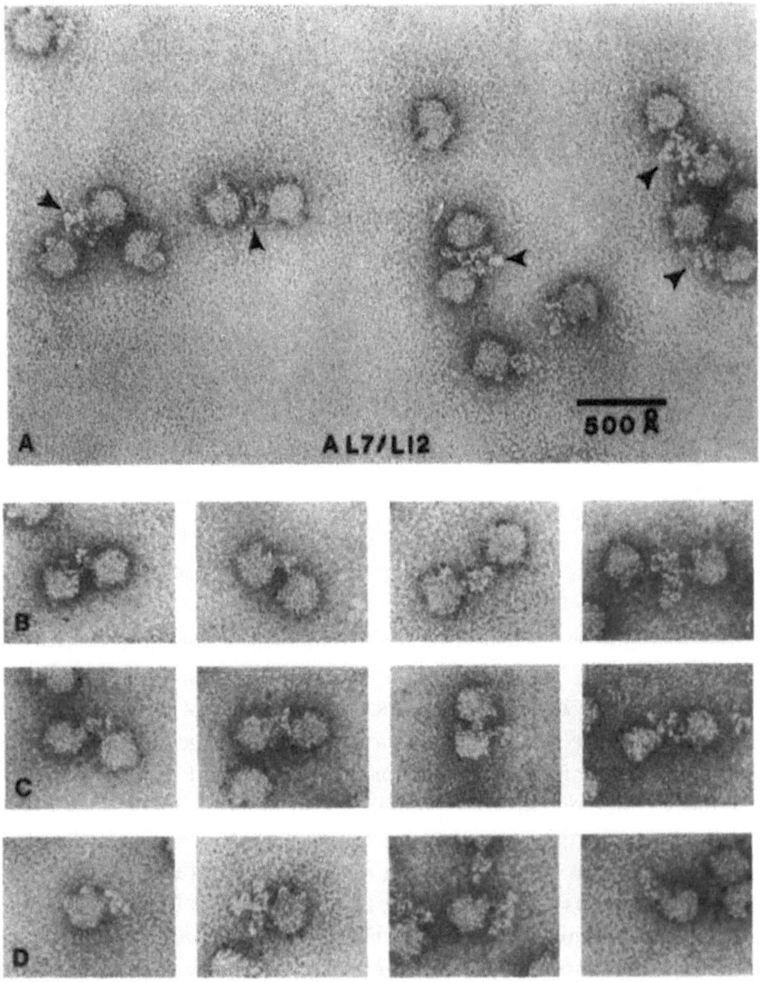

Fig. 22. A – D. Electron micrographs of large subunits reacted with AL7/L12 antibodies. The antibodies attach to a single region of the subunit, on the elongated rod-like appendage. (A) A field showing pairs of subunits linked by single, or multiple, antibodies (indicated by arrows). (B) A gallery of pairs in which a quasi-symmetric profile is at the left of each frame. (C) A gallery of pairs with the subunit at the left in the asymmetric profile. (D) A gallery of monomers with antibodies attached

The final bit of information concerning the localization and orientation of the initiator tRNA came from studies of the structure of *E. coli* small subunits, large subunits, and monomeric ribosomes, and in particular, from determining the orientation of the small subunit with respect to the large subunit in the 70 S monomeric ribosome (LAKE, 1976). Because of the larger size of the 70 S ribosome, the double layer contrasting technique tends to flatten them and make their profiles difficult to interpret. As a result, the single layer technique was used to observe the 70 S structure. A gallery of ribosomes is shown in Figure 24 and a diagram of the orientations of both subunits in these projections is shown in Figure 25. One of the projections is referred to as the overlap view (at the left in Fig. 25) since the profiles of the large and small subunits overlap (see Fig. 24 A and B). Images of large subunits (the

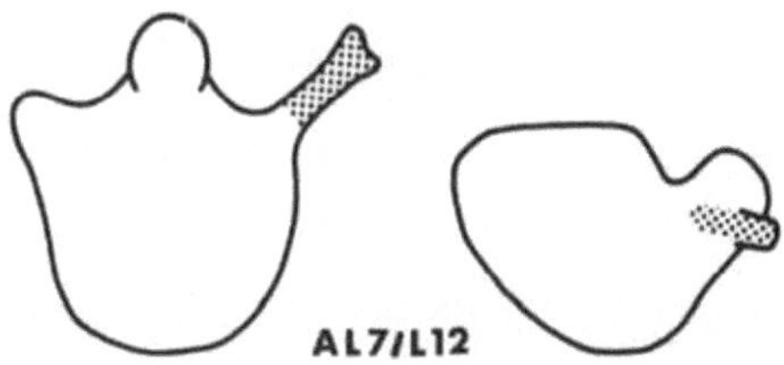

Fig. 23. A diagrammatic representation of the site of attachment of antibodies to proteins L7/L12. The site is represented by stippling in the quasi-symmetric view (*left*) and the asymmetric view (*right*)

three frames at the left) and of small subunits (the two frames at the right) in the same orientations as they have in the monomeric ribosome, in the orientation of the subunits in the overlap view, are shown in Figure 24 C. The other projection, the nonoverlap view (at the right in Fig. 25) corresponds to minimum overlap of the large and small subunit profiles (see Fig. 24 D and E). The orientations of large subunits (the two left frames) and of small subunits (the three right frames) in this view are illustrated in Figure 24 F.

Direct electron-microscopic visualization of monomeric ribosomes in these experiments provided evidence suggesting that the platform of the small subunit is located on the side of the small subunit facing the large subunit as shown in Figure 25. The interface, however, is a complicated region of overlapping densities, and the possibility existed that some other structural feature that mimics the appearance of the platform could have resulted in an incorrect determination of the small subunit orientation. Because *exact* knowledge of the orientation of the small subunit in the monomer is essential to understanding ribosome function, antibody-labeling experiments were undertaken to determine if the platform of the 30 S subunit faced the interface or not.

A field of a mixture of 70 S, 50 S, and 30 S subunits reacted with AS13 antibodies is shown in Figure 26. Antibodies against protein S13 were used since they react at a single region of the small subunit, they do not cross react with 50 S or other 30 S proteins (L. KAHAN, personal communication), they formed for approximately one-half of the 30S proteins. These locations

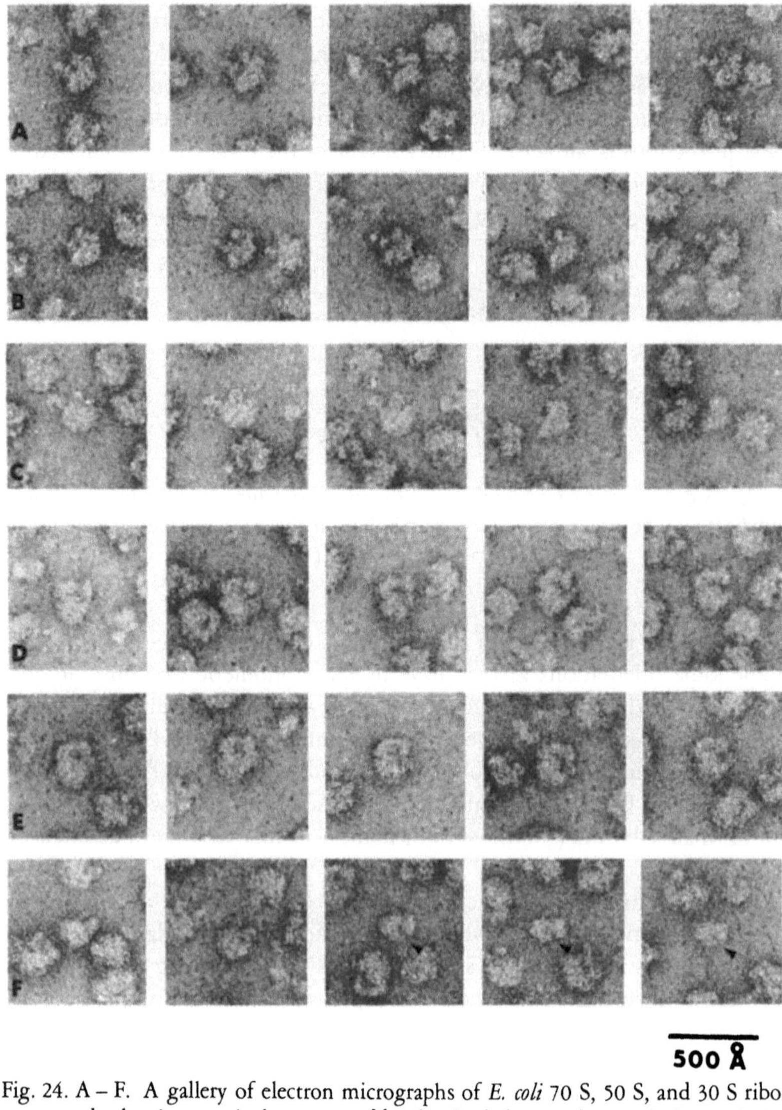

Fig. 24. A – F. A gallery of electron micrographs of *E. coli* 70 S, 50 S, and 30 S ribosomes and subunits negatively contrasted by the single layer carbon technique. (A, B) 70 S ribosomes in the overlap view, with the rod-like appendage of the 50 S subunit to the left of the central protuberance. (C) 50 S subunits (first three frames) and 30 S subunits (last two frames) in the views that correspond to their orientations in the ribosomes in rows (A) and (B). (D, E) 70 S ribosomes in the nonoverlap orientation. (F) 50 S subunits (first two frames) and 30 S subunits (last three frames) in the views that correspond to their orientations in the ribosomes in rows (D) and (E). The small arrowheads in the last three frames indicate the platforms of the small subunit. From LAKE (1976)

are on the same side of the subunit as the platform in the intermediate (+50°) view, and they bind in the upper one-third of the subunit, i.e., near the outside edges of the monomeric ribosome in the nonoverlap view so that the antibody can be seen in a region where it does not overlap the ribosome image. The field has been negatively contrasted by the double carbon method, which obscures structural details of 70 S ribosomes but permits the antibody molecules to be visualized. The arrows indicate pairs of ribosomes and

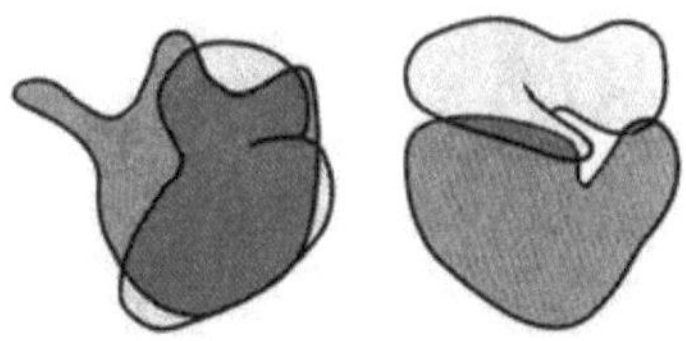

Fig. 25. Diagrammatic representations of the overlap (*left*) and nonoverlap (*right*) views of the 70 S ribosome. The large subunit is indicated by the darker lines that slant from lower left to upper right and the small subunit by the lighter lines slanting from lower right to upper left

subunits linked by antibodies. From left to right they indicate 70 S – 30 S, 70 S – 70 S, 70 S – 30 S, and 30 S – 30 S couples. The 70 S – 30 S couples were the most useful for identification since the site of antibody attachment to the 30 S and the site of antibody attachment in the 70 S profile can both be verified. Figure 26 shows images of small subunits reacted with antibodies directed against protein S13 illustrating the site of attachment of this antibody. Two AS13 antibody molecules are simultaneously bound to both subunits of a pair in about 70% of the images, indicating exposure of multiple antigenic determinants arranged such that binding of AS13 antibody at one determinant does not hinder binding at the other due either to the distance between the sites or to their relative orientation. In the first two frames the platform is seen in projection to the right of the center line of the small subunit, in the second two frames, it is to the left, and a diagrammatic interpretation of the fourth frame is shown in the final frame of 26 B. A gallery of 70 S – 30 S pairs is shown in Figure 26 C and D. In all 70 S profiles the small subunit is oriented above the large subunit with the large subunit–small subunit interface approximately horizontal. In Figure 26 C the one-third portion of the small subunit within the 70 S profile is at the right and in Figure 26 D it is at the left. The image in the fourth frame of both 26 C and 26 D is diagrammatically interpreted in the last frame of each row. Although the 30 S platform cannot be discerned in these images, the antibody clearly attaches at the interface between large and small subunits, confirming that the small subunit is oriented with the platform facing the large subunit as shown in Figure 27.

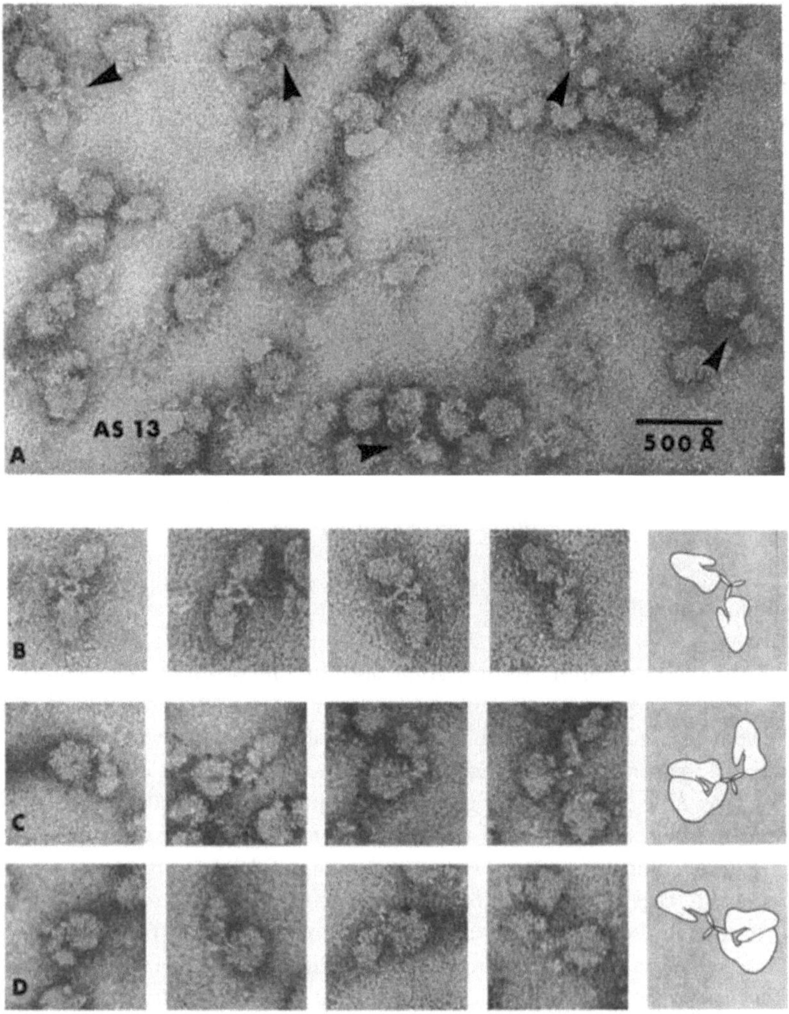

Fig. 26. A – D. Electron micrographs of 70 S, 50 S, and 30 S ribosomes and subunits reacted with antibodies directed against small subunit protein S13. (A) A field showing 70 S – 30 S, 70 S – 70 S, 70 S – 30 S, and 30 S – 30 S pairs indicated by the arrows, from left to right, respectively. (B) Small subunit pairs linked by anti-S13 antibodies, shown diagrammatically in the last frame. (C) 70 S – 30 S pairs linked by anti-S13 antibodies with the small subunit to the right, as shown diagrammatically in the last frame. (D) 70 S – 30 S pairs linked by anti-S13 antibodies with the small subunit to the left, as shown diagrammatically in the last frame. From LAKE (1976)

The importance of determining the orientation of the small subunit in the monomeric ribosome is that this information can be used to orient the initiator tRNA binding site on the ribosome. If the tRNA molecule does not undergo gross conformational changes during the protein synthesis microcycle, the structural consideration that the platform of the small subunit faces the large subunit, together with the localization of the initiator tRNA in the cleft, suggests that the CCA end of the tRNA is constrained to a limit-

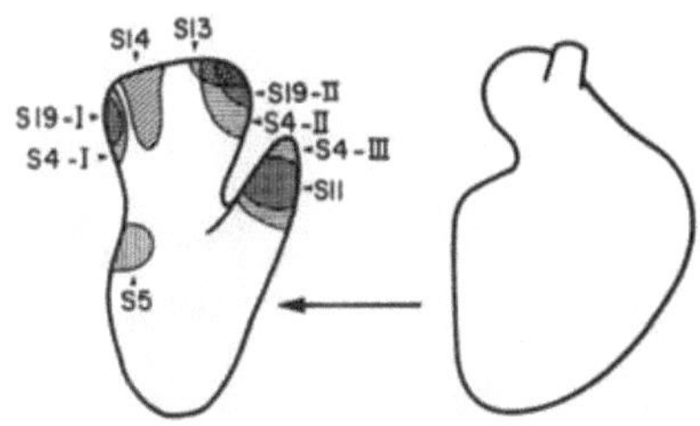

Fig. 27. A diagram showing the orientation of the small subunit and the locations of small subunit proteins with respect to the large subunit

ed area of the large subunit. In the asymmetric view of the large subunit (corresponding to the orientation of the large subunit in the overlap view of the 70 S ribosome) this area comprises the notch and the region of the central protuberance that faces the subunit interface as shown by the stippled region at the right in Figure 28. In the quasi-symmetric view (corresponding to the orientation of the large subunit in the nonoverlap view of the monomeric ribosome), this region consists of the part of the large subunit indicated by the stippling in Figure 28. The area indicated is approximate; however, a flexing of the tRNA at the TψC region during protein synthesis (as has been suggested by RICHTER et al., 1973) would not greatly modify the stippled region, provided that the anticodon stem of the tRNA remained positioned in the cleft. Thus as a result of antibody-labeling studies on the ribo-

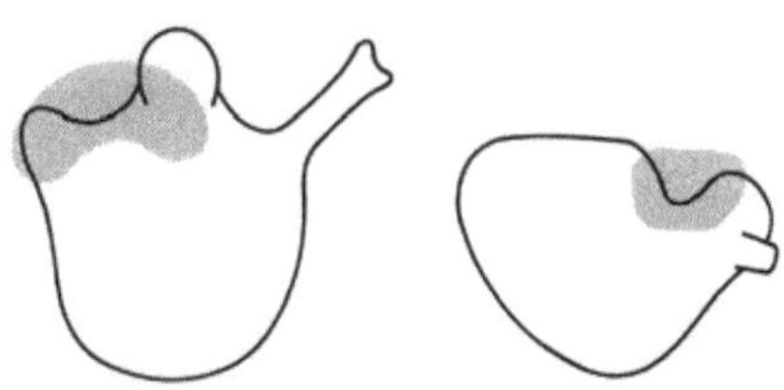

Fig. 28. The large subunit is shown diagrammatically in the quasi-symmetric (*left*) and in the asymmetric view (*right*). The stippling indicates the possible region of contact of the amino acid acceptor end of the tRNA assuming the location for the codon-anticodon interaction suggested by LAKE and KAHAN (1975)

some, the initiator tRNA binding site is beginning to be localized and implies that either the A or the P site must be oriented approximately as shown in Figure 29.

Another important aspect of the small subunit orientation it that it defined two regions of the subunit: an interface side and an external, or cytoplas-

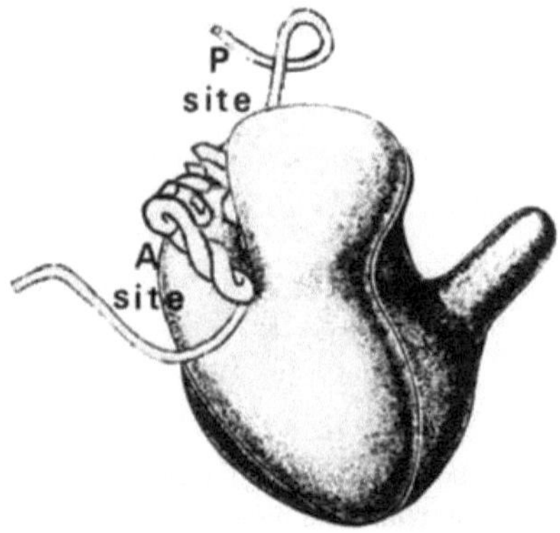

Fig. 29. Model of the monomeric ribosome showing A and P tRNA binding sites suggested by the protein locations

mic, side. In order for an A site tRNA or a P site tRNA to simultaneously contact the codon-anticodon on the small subunit and the peptidyl transferase on the large subunit, it must be bound across the interface region. Yet some proteins implicated in tRNA binding are located on the external surface of the ribosome in a region distant from the presumed A and P binding sites. This is illustrated in Figure 30 using a specific protein, S5, and the IF3 binding site. Protein S5 has been implicated in tRNA binding and codon recognition because some *E. coli* revertants from streptomycin-dependent

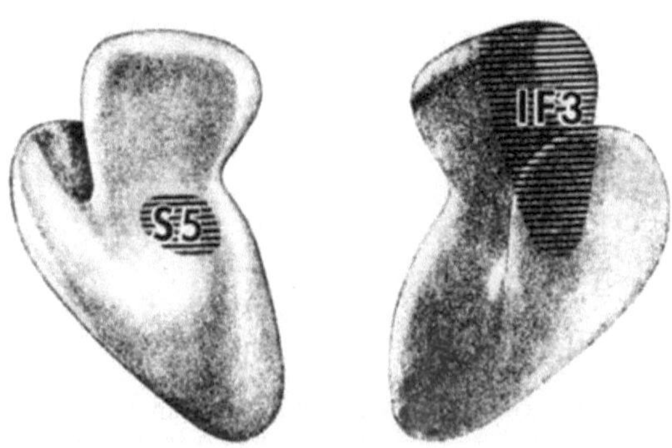

Fig. 30. Locations of protein S5 (*left*) and IF3 (*right*) on the surface of the small subunit

strains have an altered S5. Protein S5 is located on the exterior side of the small subunit and the IF3-binding site is located on the interface side of the subunit, i.e. the side adjacent to the 50S subunit.

This suggested that the elongation cycle needed to be reexamined since a tRNA cannot simultaneously contact the messenger RNA and the peptidyl

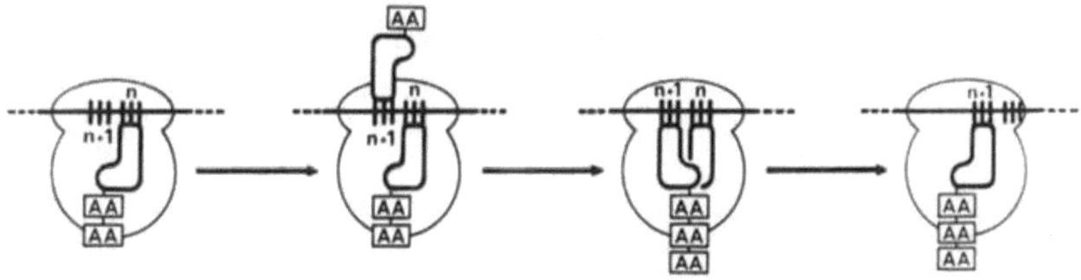

Fig. 31. The elongation cycle of protein synthesis

transferase and still contact S5. As a result the entirely new concept of the recognition (R) site was developed (LAKE, 1977).

Current knowledge of the elongation cycle is summarized in Figure 31. A cycle of elongation has been completed and the n+1 codon is available to be read by the correct tRNA in Figure 31 a. In the first step of the next elongation cycle (codon recognition, Figure 31 b) the codon is recognized and the correct aminoacyl-tRNA-EF-Tu · GTP complex is bound by the ribosome. In the second step (peptide bond formation, Figure 31 c) the GTP is cleaved to GDP and P_i, the EF-Tu · GDP complex and P_i are released from the ribosome, the peptide bond is completed, and the n+1 tRNA occupies the A site. The third event, translocation, refers to a switching of the peptidyl-tRNA from the A site (Figure 31 c) to the P site (Fig. 31 d). This step utilizes EF-G and the cleavage of GTP and results in the release of EF-G, GDP, P_i, and the uncharged tRNA.

Each of the three states shown in Figure 31 b, c, and d is a functionally distinct state (and therefore physically distinct state) of the elongation cycle. During switching from the state in Figure 31 b to the state in Figure 31 c, one GTP is converted to GDP and P_i; likewise, during switching from the state in Figure 31 c to the state in Figure 31 d, one GTP is converted to GDP and P_i. The three states are functionally distinguishable according to the following criteria. If peptidyl-tRNA is reactive with puromycin it is said to be in the P site (Fig. 31 d) and if peptidyl-tRNA is not reactive with puromycin then it is in the A site (Fig. 31 c). In a similar way, aminoacyl-tRNA in the presence of both EF-Tu and the nonhydrolyzable analog of GTP, GMPPCP, bound to the ribosome during codon recognition (Fig. 31 b) can be distinguished because peptidyl transfer cannot occur. Just as the A and P tRNA-binding sites refer to tRNAs bound to the ribosome in well-defined physical states, the term R (or recognition) site describes the distinct physical state of the aminoacyl-tRNA when bound to the ribosome during initial codon recognition.

The approximate location suggested for the P site is shown occupied by a tRNA in Figure 32 b. Similarly, in order for the A site tRNA to contact the messenger RNA at the codon-anticodon interaction site and to simultaneously contact the large subunit in the vicinity of the peptidyl transferase, it might be positioned as shown in Figure 32 b. The important fact to note, however, is that because both the A and the P site tRNAs are on the 30S – 50S interface side of the small subunit (in order to have their acceptor stems near the peptidyl transferase) they cannot be near protein S5, which is located on the external surface of the small subunit (Fig. 30). Hence, because the cluster of proteins on the external surface of the small subunit contains proteins known to affect codon recognition, and because this region cannot be a part of the A or P tRNA-binding sites, it is likely that it is part of the R site.

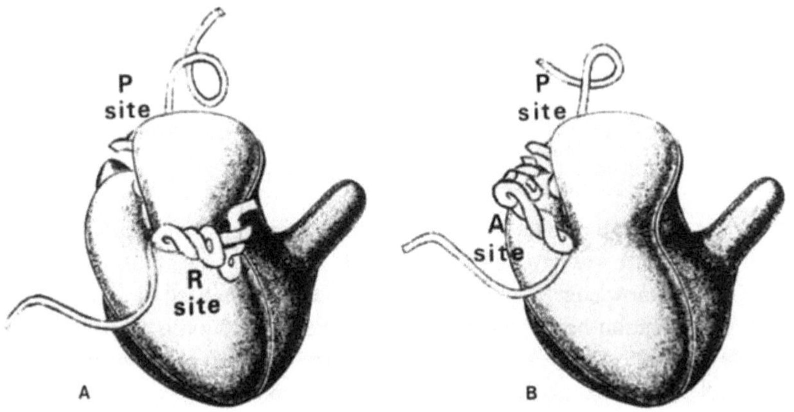

Fig. 32 A and B. Suggested locations of the tRNA binding sites on the ribosome

The tRNA is shown bound to this region in the vicinity of S5 in Figure 32 a. The amino acid acceptor end of the tRNA is oriented near the cluster of proteins S14, S19-I, and S4-I (not labeled). Because EF-Tu is thought to bind to the acceptor region of the tRNA, EF-Tu is also probably near the cluster of proteins S14, S19-I, and S4-I and near the elongated rod-like appendage of the large subunit where L7 and L12, large subunit proteins required for EF-Tu dependent GTP hydrolysis, were localized.

In summary the existence of the R tRNA binding site is necessary, given the current understanding of the elongation cycle. Just as the A and the P tRNA sites describe tRNAs bound to the ribosome in well-defined physical states, the R site describes the functionally and physically distinct state of the aminoacyl tRNA bound to the ribosome during initial codon recognition.

The concept of the R site is supported by the most basic ideas concerning the elongation cycle and obivously its existence is not dependent on the correctness of the details of the location of the site. The concept of the R site was generated indirectly by the locations of the proteins determined from antibody mapping studies. Hence we find that structural studies are advancing our understanding of ribosome function and leading to new concepts of protein synthesis.

Antibody labeling is now providing structural information about the three-dimensional distributions of ribosomal proteins that only a few years ago, was practically unthinkable. Hopefully, in the next years, as a result of combining the results of functional studies with structural studies, a structural understanding of the molecular events occurring during protein synthesis will emerge.

Acknowledgements. I especially thank my collaborators in these antibody-labeling experiments, Drs. L. KAHAN and M. NOMURA, for providing helpful advice and suggestions. I also thank Dr. L. KAHAN for reading the manuscript and Dr. M. NOMURA for providing a copy of the assembly map.

References

Bode, U., Lutter, L. C., Stöffler, G.: Proteins S14 and S19 are near-neighbors in the *E. coli* ribosome. FEBS Lett. **45**, 232 – 236 (1974)

Bollen, A., Kahan, L., Heimark, R. L., Cozzone, A., Traut, R. R., Hershey, J. W. B.: Cross-linking of initiation factor IF-2 to *E. coli* 30 S ribosomal proteins with dimethylsuberimidate. J. Biol. Chem. **250**, 4310 – 4314 (1975)

Brenner, S., Horne, R. W.: A negative staining method for high resolution electron microscopy of viruses. Biochim. Biophys. Acta **34**, 103 – 110 (1959)

Bruskov, V. I., Kieselev, N. A.: Electron microscope study of the structure of *E. coli* ribosomes and CM-like particles. J. Mol. Biol. **37**, 367 – 377 (1968)

Eiserling, F., Levin, J. G., Byrne, R., Karlsson, U., Nirenberg, M. W., Sjostrand, F. S.: Polysomes and DNA-dependent amino acid incorporation in *E. coli* extracts. J. Mol. Biol. **10**, 536 – 539 (1964)

Gorini, L.: Ribosomal discrimination of tRNAs. Nature (New Biol.) **234**, 261 – 264 (1971)

Hall, C. E., Slayter, H. S.: Electron microscopy of ribonucleoprotein particles from *E. coli*. J. Mol. Biol. **1**, 329 – 332 (1959)

Hart, R. G.: Electron microscopy of 50 S ribosomes of *E. coli*. Biochim. Biophys. Acta **60**, 629 – 637 (1962)

Hart, R. G.: Surface features of the 50S ribosomal component of *E. coli*. Proc. Natl. Acad. Sci. U. S. **53**, 1415 – 1420 (1965)

Held, W. A., Ballow, B., Mizushima, S., Nomura, M.: Assembly mapping of 30 S ribosomal proteins from *E. coli*: Further studies. J. Biol. Chem. **249**, 3103 – 3111 (1974)

Higo, K., Held, W., Kahan, L., Nomura, M.: Functional correspondence between 30 S ribosomal proteins of *E. coli* and *B. stearothermophilus.* Proc. Natl. Acad. Sci. U. S. **70**, 944 – 948 (1973)

Higo, K., Loertscher, K.: Amino terminal sequence of some *E. coli* 30 S ribosomal proteins and functionally corresponding *B. stearothermophilus* ribosomal proteins. J. Bacteriol. **118**, 180 – 186 (1974)

Hill, W. E., Thompson, J. D., Anderegg, J. W.: X-ray scattering study of ribosomes from *Escherichia coli.* J. Mol. Biol. **44**, 89 – 102 (1969)

Huxley, H. E., Zubay, G.: Electron microscope observations on the structure of microsomal particles from *E. coli.* J. Mol. Biol. **2**, 10 – 18 (1960)

Kabat, E. A.: Structural concepts in immunology and immunochemistry. New York-Toronto-London: Holt, Rinehart and Winston (1968)

Kahan, L., Held, W. A., Nomura, M.: Effects of aminoethylation of cysteine residues on the structural and functional activities of the *E. coli* 30 S ribosomal proteins. J. Mol. Biol. **88**, 797 – 808 (1974)

Lake, J. A.: Aminoacyl-tRNA binding at the recognition site is the first step of the elongation cycle of protein synthesis. Proc. Natl. Acad. Sci. U.S.A. **74**, 1903 – 1907 (1977)

Lake, J. A.: Ribosome structure determined by electron microscopy of *E. coli* small subunits, large subunits and monomeric ribosomes. J. Mol. Biol. **105**, 131 – 160 (1976)

Lake, J. A., Kahan, L.: Ribosomal proteins S5, S11, S13 and S19 localized by electron microscopy of antibody-labeled subunits. J. Mol. Biol. **99**, 631 – 644 (1975)

Lake, J. A., Leonard, K. R.: Structure and protein distribution for the capsid of *Caulobacter crescentus* bacteriophage ΦCbK. J. Mol. Biol. **86**, 499 – 518 (1974)

Lake, J. A., Nonomura, Y., Sabatini, D. D.: Ribosome structure as studied by electron microscopy. In: Ribosomes. Nomura, M., Tissieres, A., Lengyel, P. (eds.). New York: Cold Spring Harbor Laboratory Press, 1974 a, pp. 543 – 557

Lake, J. A., Pendergast, M., Kahan, L., Nomura, M.: Ribosome structure: three dimensional distribution of proteins S14 and S4. J. Supramol. Struc. **2**, 189 – 195 (1974 b)

Lake, J. A., Pendergast, M., Kahan, L., Nomura, M.: Localization of *E. coli* ribosomal proteins S4 and S14 by electron microscopy of antibody-labeled subunits. Proc. Natl. Acad. Sci. U. S. **71**, 4688 – 4692 (1974 c)

Leonard, K. R., Kleinschmidt, A. K., Agabian-Kashishian, N., Shapiro, L., Maizel, J. W., Jr.: Structural studies on the capsid of *Caulobacter crescentus* bacteriophage ΦCbK. J. Mol. Biol. **71**, 201 – 216 (1972)

Leonard, K. R., Kleinschmidt, A. K., Lake, J. A.: *Caulobacter crescentus* bacteriophage ΦCbK: structure and *in vitro* self-assembly of the tail. J. Mol. Biol. **81**, 349 – 365 (1973)

Lubin, M.: Observations on the shape of the 50 S ribosomal subunit. Proc. Natl. Acad. Sci. U. S. **61**, 1454 – 1461 (1968)

Lutter, L. C., Kurland, C. G.: Reconstitution of active ribosomes with cross-linked proteins. Nature (New Biol.) **243**, 15 – 18 (1973)

Mizushima, S., Nomura, M.: Assembly mapping of 30 S ribosomal proteins from *E. coli.* Nature (London) **266**, 1214 – 1218 (1970)

Nomura, M., Mizushima, S., Ozaki, M., Traub, P., Lowry, C. V.: Structure and function of ribosomes and their molecular components. Cold Spring Harbor Symp. Quant. Biol. **34**, 49 – 61 (1969)

Nomura, M., Tissieres, A., Lengyel, P. (eds.): Ribosomes, New York: Cold Spring Harbor Laboratory Press (1974)

Reinbolt, J., Schiltz, E.: The primary structure of ribosomal protein S4 from *E. coli.* FEBS Lett. **36**, 250 – 253 (1973)

Richter, D., Erdmann, V. A., Sprinzl, M.: Specific recognition of GTψC loop (loop IV) of tRNA by 50 S ribosomal subunits from *E. coli.* Nature (New Biol.) **246**, 132 – 135 (1973)

Sommer, A., Traut, R. R.: Diagonal polyacrylamide-dodecyl sulfate gel electrophoresis for the identification of ribosomal proteins crosslinked with methyl-4-mercaptobutyrimidate. Proc. Natl. Acad. Sci. U. S. **71**, 3946 – 3950 (1974)

Stöffler, G., Wittmann, H. G.: Sequence differences of *E. coli* 30 S ribosomal proteins as determined by immunochemical methods. Proc. Natl. Acad. Sci. U. S. **68**, 2283 – 2287 (1971)

Sun, T-T., Bollen, A., Kahan, L., Traut, R. R.: Topography of ribosomal proteins of the *E. coli* 30 S subunit as studied with the reversible crosslinking reagent methyl-4-mercaptobutyrimidate. Biochemistry **13**, 2334 – 2340 (1974)

Traub, P., Nomura, M.: Structure and function of *E. coli* ribosomes. Reconstitution of functionally active 30 S ribosomal particles from RNA and proteins. Proc. Natl. Acad. Sci. U. S. **59**, 777 – 781 (1968)

Valentine, R. C., Green, M.: Electron microscopy of an antibody-hapten complex. J. Mol. Biol. **27**, 615 – 617 (1967)

Valentine, R. C., Shapiro, B. M., Stadtman, E. R.: Regulation of glutamine synthetase XII. Electron microscopy of the enzyme from *E. coli.* Biochemistry **7**, 2143 – 2152 (1968)

Vasiliev, V. D.: Electron microscopy study of 70 S ribosomes of *E. coli.* FEBS Lett. **14**, 203 – 205 (1971)

Vasiliev, V. D.: Morphology of the ribosomal 30 S subparticle according to electron microscopic data. Acta Biol. Med. Germ. **33**, 779 – 793 (1974)

Wabl, M. R.: Electron microscopic localization of two proteins on the surface of the 50 S ribosomal subunit of *E. coli* using specific antibody markers. J. Mol. Biol. **84**, 241 – 247 (1974)

Wabl, M. R., Barends, P. J., Nanninga, N.: Tilting experiments with negatively stained *E. coli* ribosomal subunits. An electron microscopic study. Cytobiologie **7**, 1 – 9 (1973 a)

Wabl, M. R., Doberer, H. G., Hoglund, S., Ljung, L.: Electron microscopic study of isolated 30 S ribosomal subunits of *E. coli.* Cytologie **7**, 111 – 115 (1973 b)

Wittmann, H. G., Wittmann-Liebold, B.: Chemical structure of bacterial ribosomal proteins. In: Ribosomes. Nomura, M., Tissieres, A., Lengeyel, P. (eds.). New York: Cold Spring Harbor Laboratory Press, 1974, pp. 115 – 140

Yanagida, M. and Ahmad-Zadeh, C.: Determination of gene product locations in bacteriophage T4 by specific antibody association. J. Mol. Biol. **51**, 411 – 421 (1970)

Electron Microscopy and Electron Diffraction Studies on Hydrated Membranes

S. W. HUI and D. F. PARSONS

A. Introduction

Despite nearly 30 years of applying electron microscopy to the problem, we are still in doubt about coarse structural features, e.g., whether the central part of the membrane is occupied only by lipid or by a mixture of lipid and protein. Since the electron-microscopic resolution available (better than 5 Å) has been adequate for some time, we can only conclude that the problem is due either to inadequate contrast, inadequate specimen preservation, or both. Until recently, most information was obtained from the cross sections and tangential sections obtained by the thin sectioning method. It is now realized that the 'railway track' appearance seen in cross section is an unreliable indicator of the internal membrane structure. As ordinarily prepared, the membrane is depleted of lipids and, in many cases, probably disorganized by the fixation and embedding technique (MORETZ et al., 1969). In thin sections only the metal stain is visible; the membrane constituents themselves are not. The method also usually fails to show the polysaccharide coat even when it is present as a thick layer (PARSONS and SUBJECK, 1972). It appears that satisfactory infiltration of cells with the usual plastic embedding media (e.g., Epon) requires rendering the membranes porous by prior extraction of the lipids (ROBERTSON and PARSONS, 1970). Thin-section embedding techniques have been improved by the use of 'inert dehydration', e.g., glycerol (PEASE, 1973), and the use of embedding materials (e.g., glycol/methacrylate/glutaraldehyde/urea mixture) that retain lipids (PEASE, 1974). Variable membrane cross-sectional appearances (from railway track to featureless) can be obtained by different staining methods, but it is noteworthy that the method has failed to show the intramembrane particles (IMP) seen so abundantly in the center of the membranes by the freeze-fracture-etch method.

The freeze-etch method has provided the most dramatic new information about the interior structure of membranes. The method suggests that all high-protein membranes contain globular proteins of 80 to 100 Å size at their center, arranged either in clusters or fairly evenly distributed about 150 Å apart. However, the failure to see such particles by any other electron-microscopic technique or by X-ray diffraction raises a serious question of a possible artifact related to quenching the tissue and freezing it at liquid nitrogen temperatures. Further experimentation with the freeze-etch and X-ray diffraction (or electron diffraction) techniques are necessary to establish the fact that abundant globular proteins occupy the center of the membrane.

Negative staining has, of course, been useful for demonstrating surface projections of membranes, although it gives a poor indication of polysaccharide structure (PARSONS and SUBJECK, 1972). The loss of fine structures and the introduction of artifacts in the air drying process are also unavoidable in this method.

Currently, electron-microscopic indications of internal membrane structure remain confused. While freeze-etching could be expected to be a more artifact-free specimen preparation method (only requiring freezing or freezing plus a cryoprotectant), another minimal artifact technique is necessary to confirm and extend its results. This article is a discussion of the use of electron microscopy (EM) and electron diffraction (ED) of wet cell membranes as a new technique to fill this gap.

Electron diffraction has seldom been applied to the study of membrane structures, despite its advantages. The main drawback has been the need to dehydrate the specimen. This process often leads to the destruction of the details in periodic or quasi-periodic structures that electron diffraction is geared to observe (MATRICARDI et al., 1972 a). There is a fair amount of evidence to show that membranes change their structure on air or vacuum drying. Striking powder ED patterns can be obtained from vacuum-dried red blood cell ghost membranes (Fig. 1). The lipid appears to have been extruded and recrystallized. It is also a common experience with pellets of membranes examined by X-ray diffraction (XD) that the initial diffuse pattern first changes to a sharper one as partial dehydration leads to better stacking of the membranes and then changes again as the membranes become disorganized by drying. So far, all published literature on the ED of membranes pointed to the inevitable fact of recrystallization of membrane components upon drying. Early works on the ED of phospholipid deposits (HURST, 1952) suggested that dried lipid bilayers were indistinguishable from dried bulk crystals. Subsequent work on dried myelin (FERNANDEZ-MORAN, 1951), retina rod (MATHEJA, 1971), brain cell membrane (KHASE and MISRA, 1973), and the plasma membranes of *Acholeplasma laidlawii* (CARLEMALM and WEISLANDER 1975) all show gross recrystallization of salt and various membrane components. These recrystallizations in the drying process make the extrapolation of the ED data extremely difficult. This fact alone made ED initially less attractive than XD in the study of membrane structure. To preserve the speci-

Fig. 1. Electron diffraction (ED) pattern of vacuum dried red blood cell ghost membranes

men during electron-microscopic and ED observation, the specimen needs to be contained in a hydration chamber that is essentially isolated from the vacuum of the electron microscope. This method will be discussed in Section B.

Alternatively, the hydrated specimen can be a frozen specimen that contains a minimum of excess water. Frozen thin sectioning is slowly developing as a practical method (CHRISTENSON, 1971; BERNHARD, 1974), but sections cut by this method still do not show the same IMP seen in freeze-etch (perhaps because of inadequate contrast). More recently a few investigators have begun to examine specimens frozen on the grid with only a thin layer of ice water (HEIDE and GRUND, 1974; TAYLOR and GLAESER, 1974), and it is clear that some possibilities exist for dark-field observation of the surfaces of frozen membranes in flattened or small vesicle form. However, frozen materials have the disadvantage over wet specimens that observation is restricted to temperatures lower than the physiologic temperature and the transition temperature of membrane lipids, as well as having additional risk of ice damage. Sugar substitution has been used to preserve the structures of hydrated protein crystal and membranes (UNWIN and HENDERSON, 1975). This method may be useful in bridging the gap between negative staining and hydration chamber methods. High resolution results similar to those obtained from wet specimens have been obtained by this method. A novel technique has been developed by BASU and PARSONS (1974) to replicate the wet cell surface by silicone monoxide evaporation. With this technique the structure of cell surface polysaccharides in a hydrated state may be observable. The images should be comparable to those obtained by back-scattered scanning electron micrographs using a hydration chamber (ROBINSON, 1975).

B. Operation of Hydration Chamber in an Electron Microscope

I. Hydration Chamber

1. Principles

The design, operation, and results have already been reviewed (JOY, 1973; PARSONS et al., 1974; PARSONS, 1974). The environmental chamber is a small compartment for the wet specimen that is either closed by thin windows or by small apertures. So far, the differentially pumped, aperture-limited chamber has been most widely used because thin windows break or leak easily. The wet grid is enclosed by two apertures large enough to admit the electron beam but small enough (75 – 100 μm) to give only a slow leak of water vapor. The wet grid rapidly comes into thermodynamic equilibrium with the enclosed reservoir of water and its vapor. The small leak of water vapor is differentially pumped out of the microscope in a second compartment that surrounds the inner one and is closed by larger apertures (200 μm). A schematic drawing is shown in Figure 2. The whole unit is thermally insulated from the microscope and can be cooled or heated by a thermoelectric device. A prototype design is described by HUI et al. (1976).

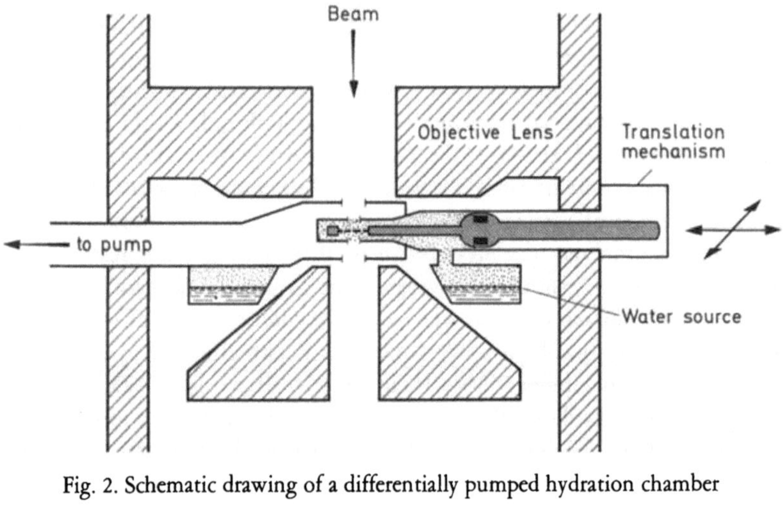

Fig. 2. Schematic drawing of a differentially pumped hydration chamber

2. Chambers for Fixed-beam Transmission Electron Microscope

Our experience to date has resulted in several general conclusions about the ease of use of different types of environmental chambers and the use of different microscope image modes, based on the construction of three modifications for a JEM 200 (200 kV) in which the chamber had to be placed above (or dipping down into) the upper objective pole piece because of lack of sufficient access ports at the objective lens gap level. This modification yields an image resolution of better than 20 Å and unrestricted wide-angle and small-angle electron diffraction. Another 'above the lens' modification (temperature controlled), built for the Siemens Ia, has operated in a particularly trouble-free manner (HUI et al., 1976). Two 'in the gap' modifications have been built: one for the U. S. Steel (Monreoville, Pa., High Voltage Electron Microscopy Laboratory under the direction of Dr. R. Fisher) 1.0 MeV microscope. The large size of the objective pole pieces, and ports at the lens gap level, made it simpler to construct a temperature-controlled (circulating ethylene glycol) model that was used to obtain about 4000 negatives of wet cells (PARSONS et al., 1972, 1974; HUI and PARSONS, in press). Side entry, retractable environmental chambers of the window type [based on ALLINSON's design, (Allinson et al., 1972)], and of the differentially pumped type (based on SWANN's design, 1972) have also been built for the AEI EM7 high voltage electron microscope. A detachable cartridge chamber for this microscope has recently been completed (PARSONS, unpublished). Another 'in the gap' modification was built for the Philips EM 300 in the STEM mode (PARSONS and LYON, 1975). In this case, difficulties were met (now partially overcome) in minaturization of the chamber for the small lens gap and maintaining the chamber aperture environment concentric with the critical STEM mode column alignment.

In general, highest resolution and ease of operation are obtained with a device in the gap of a high-voltage electron microscope (HVEM) although devices can be constructed for the lens gap of 100 kV instruments. Probably, it is best to replace a separated drive for centering the hydration stage by exact machining based on making the apertures of the chamber concentric with the bore of one-half of the objective lens pole piece, as has been done by SWANN (1972). Centering of the apertures on the beam can be obtained by a slight misalignment of one-half of the objective lens. This makes a separate x-y translation for the hydration stage unnecessary and removes a common cause of loss of resolution associated with vibration of the stage plus specimen holder.

The resolution depends on the mechanical stability of the stage, the mass thickness of the environmental gas and water vapor (dependent on the thickness of the two compartments of the chamber and the pump speed available to remove escaping gases), and the mass thickness of the wet specimen. With the HVEM (particularly in the scanning mode) the high penetration enables high gas environment pressures (up to 1 atm of He/H_2O vapor at 1.0 MeV) and the use of thicker wet specimens. However, our experience to date indicates that wet cell nuclei are not adequately penetrated at 1.0 MeV, and a scanning attachment should be used to increase the effective penetration to see nuclear detail.

3. *Chambers for Scanning Electron Microscope*

Recently, ROBINSON (1975) gave an interesting account of simple equipment for obtaining backscatter images from wet specimens. Charging is no longer a problem with wet specimens, and coating of the specimens with a conductor has not been required. Uncoated images of wool fibers partly immersed in a thin layer of water also gave adequate contrast of fiber details without coating. The technique is obviously of interest for studying the surface of wet cells. Hopefully, sufficient resolution can be obtained to see fine cell processes and aggregations or domains in the polysaccharide coat. To reveal these detailed surface structures, all but a very thin layer of water must be removed from the hydrophilic cell surface (ROBINSON, 1976).

We have already reviewed the use of environmental chambers in the STEM (PARSONS et al., 1974; PARSONS 1974). The transmission scanning mode is particularly useful for penetrating thick wet objects (PARSONS and LYON, 1975) and may be the best approach for enhancing the contrast of wet unstained objects since the elastic and inelastic scatter signals can be electronically manipulated and analog contrast enhancement is readily obtained with the display electronics.

II. Preparation of Wet Membrane Specimens

The main objective is to obtain preparations in which the excess water is reduced to a minimum. In some cases, lipids or membranes can be placed on a holey support as a thin film or as a single membrane. This is the most favorable case since the diffuse scattering produced by the extraneous water is minimized. The next most favorable case is flattened vesicles (e.g., some types of Golgi apparatus) and small vesicles. Large vesicles (1 μm or more) even if penetrated adequately give too much diffuse scatter to be useful. Obtaining a variety of cell membranes in a suitable form is a problem since most membranes when ruptured reseal into spherical vesicles. However, with sufficient care, large sheets of plasma membranes can be obtained by gently disrupting the cell either osmotically (WARLEY and COOK, 1973) or mechanically (EMMELOT et al., 1964). Certain membranes such as the purple membrane of halobacterium (STOECKENIUS and ROWEN, 1967) and cytochrome oxidase membrane at low lipid content (VAIL and RILEY, 1974) occur naturally in flattened sheet form. These membranes would flatten on the grid by the surface tension of water when excess water was slowly removed. Large membrane vesicles such as erythrocyte ghosts (HUI, unpublished) may also flatten in the same manner. General experimental methods of converting resealable membranes to a flattened, single membrane are technically feasible but have not yet been worked out. Frequently, membranes rupture and remain extended on a water surface. If suitably charged substrates can be found, with high membrane affinity, the extended membranes can probably be maintained extended on the grid. Another possibility is to bring the grid up through the surface of extended membranes in such a way that excess water is immediately wiped off and surface tension is able to maintain the membranes in an extended form. Unsupported artificial and reconstituted membranes may be formed on grid holes by a method developed by HUI et al. (1974). A fine-mesh (1000 or 1500) grid is dipped from air to water or is pulled from water to air through a lipid monolayer spread on a Langmuir trough. A bilayer or an inverted bilayer (soap film) will thus form. The bilayer is stable under saturated water vapor pressure. Langmuir-Blodgett multilayers may also form on a formvar coat grid by repeated dipping (HUI, 1976 a).

The foregoing operations are carried out in a 100% humidity glove box that contains an interference microscope for adjusting the thickness of the wet specimen. It has been found that if the specimen holder and box are of uniform temperature, very thin layers less than 1000 Å thick can be formed and remain stable against evaporation. An important general point is that our combined fixed beam high voltage electron microscope work, our recent wet replica work (BASU and PARSONS, 1974), and the work of ROBINSON (1975) on backscatter scanning images of wet biological objects, indicate that water can be absorbed off the surfaces of cell membranes leaving only a very thin layer of bound water. This is shown by the visibility of projecting surface structures of only a few hundred Ångstrom size (or less). If a thick layer of bound water on the cell surface persisted, surface details would be obscured and the transmission image of internal structures obscured by extraneous scattering by the water. The preparation is placed inside a wetted closed tube for transfer to the microscope. The environmental chamber in the microscope is filled with up to 1 atmosphere of humified air and the wet specimen inserted. The humid air supply is closed off and the chamber pressure rapidly falls to the vapor pressure of water associated with the common temperature of the specimen and the inner water reservoir. The manometer reading is checked against the value expected from water vapor pressure tables for the temperature read from the thermocouple attached to the specimen grid. Agreement indicates that there are no leaks of the chamber to the atmosphere or to the high vacuum of the microscope. A more critical test is to check the electron diffraction pattern of ox liver catalase crystals since the pattern is irreversibly lost if the equilibrium water vapor pressure falls below 90% (MATRICARDI et al., 1972 a). The temperature of the whole ensemble of chamber and its water reservoir can be changed by means of a thermoelectric device. Water vapor equilibrium is obtained rapidly. If it is desired to do experiments with undersaturated water vapor, the inner water reservoir is emptied and a connection made to an external water reservoir that can be cooled to give the desired water vapor pressure.

Since the low contrast biological specimen itself and not the staining or replicating material is under the electron beam radiation, care must be taken to reduce radiation damage to the specimen. Small condenser apertures (30 μm or less) and strong first condenser lens excitation are used to reduce the radiation dose. Beam damage is minimized by photography on sensitive X-ray film (MATRICARDI et al., 1972 b) or by use of an image intensifier. Dosage is measured by a Faraday cage (TURNER et al., 1975).

C. Electron Microscopy

There are several difficulties in the imaging of wet, unstained specimens. The most obvious ones are the lack of contrast and the specimen susceptibility to radiation damage. The former results from the lack of staining and the increasing background due to the thin water layer. Several techniques have been developed to overcome these difficulties.

I. Dark Field

Dark field is a well-known technique to increase the contrast of electron images (HENKELMAN and OTTENSMEYER, 1969). It is most useful on very thin specimens where multiple scattering is low. In membrane specimens, innate contrast is provided by the difference in mass thickness between protein, glycoprotein, glycolipid, and lipid molecules. A simple calculation shows that the mass thickness difference resulting from a typical spherical protein molecule completely embedded in a phospholipid bilayer provides a contrast of 25% from the lipid background. With the increase of background from the superimposing layers of polysaccharide coat, this contrast would be less. The dark-field technique has to be pushed hard to achieve any observable contrast. So far, no irregular structure in unstained natural membrane has been imaged at the molecular level. However, the redundant information in regularly repeating structure may be utilized in the low dose imaging with the aid of signal averaging techniques. The image of the purple membrane of *Halobacterium halobium* was obtained in this manner to a resolution of 7 Å. (UNWIN and

HENDERSON, 1975). Metal ions in fatty acid salts may be utilized to enhance the contrast of images of the otherwise unstained lipid multilayers (WALDBILLIG et al., 1976). Arrays of absorbed ferritin molecules in lipid monolayers have also been imaged (FROMBERG, 1971).

A way to enhance contrast of a periodic specimen is selective reflection contrast, a method well known to materials scientists. A small number of reflections in the ED pattern are selected to form a low resolution bright- or dark-field image depending on whether the unscattered beam is included. Crystallites or 'grains' of different spacings and orientations stand out against the remainder of the field when the associated reflections are chosen to form an image. This method has been applied to study the domain structure of phospholipid bilayers (HUI and PARSONS, 1975). Selective reflection dark-field electron microscopy was carried out by placing a selection aperture in the back focal plane of the objective lens. For domain contrast enhancement, only the (100) reflections of certain solidus domains in a dipalmitoyl lecithin (DPPC) bilayer were transmitted by the aperture. The dark-field image formed by the transmitted pattern showed remarkable contrast between differently oriented domains. The domain size of DPPC bilayers measured by this method were typically several micrometers wide, as shown in Figure 3. No domain structure was observed at temperatures above the phase transition (41.5° C). In bilayers of an equimolar mixture of dipalmitoyl and dilauroyl le-

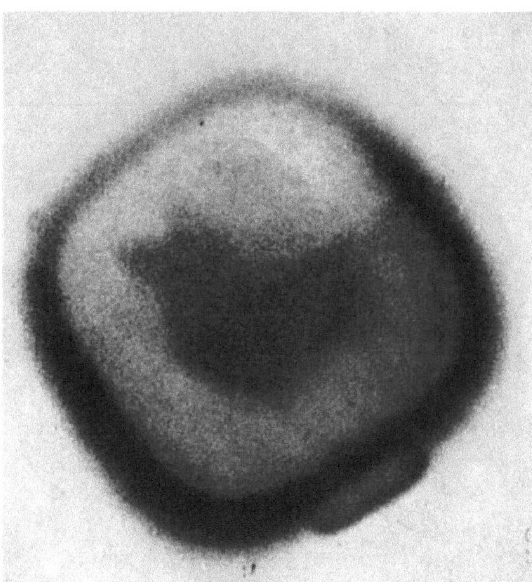

Fig. 3. Dark field micrographs of a dipalmitoyl lecithin bilayer at 17° C showing two domains (magnification ×6600)

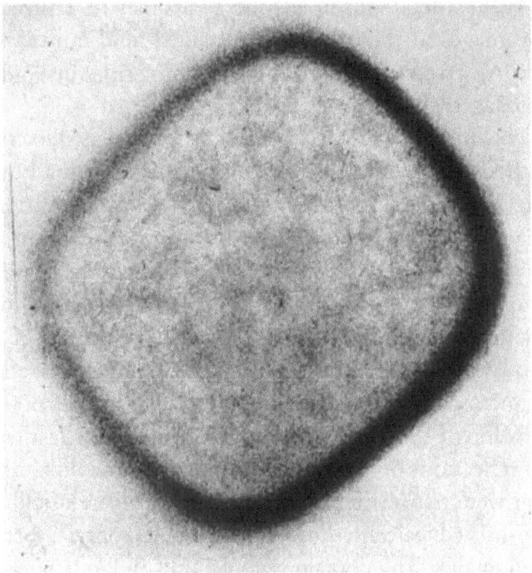

Fig. 4. Dark field micrographs of a bilayer of an equimolar mixture of dilauroyl and dipalmitoyl lecithin at 10° C. Domains of separated phases are visible (×6600)

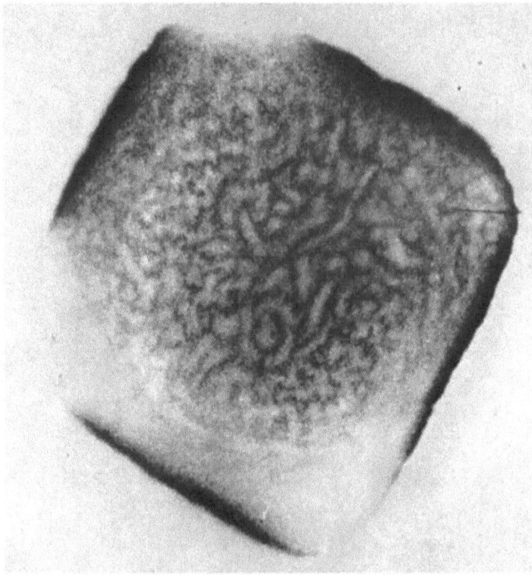

Fig. 5. Dark field micrographs of a bilayer of cholesterol-dipalmitoyl lecithin mixture at a molar ratio of 0.65. The pattern was taken at 11° C (×6600)

cithin, low contrast domains typically 300 nm wide (Fig. 4) were observed at temperatures between the higher and the lower endothermic peaks in the calorimetric scan of the mixture (D. PAPAHADJOPOULOS, private communication). The low contrast is probably due to the imperfect masking (by the selection aperture) of the diffuse diffraction pattern produced by some liquidus domains. The domain sizes measured here agree with the striated areas observed in freeze-fracture experiments (VERVERGAERT et al., 1973). In mixtures of DPPC and cholesterol, a sharp diffraction ring corresponding to the packing of pure palmitic acyl chains remained visible until the molar cholesterol content reached 1 : 1. It was therefore possible to segregate the pure DPPC reflections from the total diffraction pattern. The dark-field electron micrograph thus formed showed ribbon-like structures in mixed DPPC-cholesterol bilayers (Fig. 5), the ribbons being less than 100 nm wide. The ribbons could not be interpreted to be the loci of moving domains during the exposure time of 5 s, since the tracks were never crossed. The ratio between light and dark domains varied with the composition of the mixture. It is possible that these domains represent areas of pure components or particular eutectic complexes. Both the ribbon pattern and the DPPC diffraction ring disappeared as the temperature was raised beyond the transition temperature of the corresponding mixture.

II. Energy Filters

Energy filters are another approach for increasing contrast and resolution of both images of wet, unstained objects and of their electron diffraction pattern. This is the case because the filter can be set to cut off most of the inelastic scatter (energy losses exceeding about 15 eV). This gives contrast relation to thickness both in bright field and dark field but also enhances the resolution by reducing the inelastic electron scattering contributing to chromatic aberration. Recently, WILSKA (1974, personal communication) made very effective use of an imaging filter fitted to a low voltage microscope (6 – 10 kV) to improve contrast and resolution on objects that are thick in relation to the acceleration voltage. Eventually, there are diminishing returns with respect to the increasing electron beam damage exposure to the specimen required when much of the specimen scatter has been subtracted out.

While it is relatively easy to design energy filters for the scanning microscope, there is also a need for an imaging filter for the fixed beam microscope and for energy filtering of electron diffraction patterns. Biological electron diffraction patterns cannot be scanned across an energy filter aperture because of the excessive beam damage that would result. Imaging energy filters have been successfully applied to 100 kV instruments by use of a Castaing filter consisting of an electrostatic mirror and a triangular magnetic prism (EL HILI, 1966; HENKELMAN and OTTENSMEYER, 1974). Since the 1 MeV acceleration voltage cannot be carried in a cable, a device that includes an electrostatic mirror cannot be used for filtering fixed-beam HVEM images or electron diffraction patterns. However, recently ROSE and PLIES (1974) designed a totally magnetic energy filter that should be applicable to HVEMs.

III. Image Intensifiers

Image intensifiers are very useful for medium resolution imaging and recording of electron diffraction patterns. We use a Westinghouse TEM 801 image intensifier. This has a maximum resolution at 1×10^{-4} foot candles of 600 lines and uses a 1 in SEC WL-30839 image tube coupled to a WL-30677 image intensifier. It is important to use fiber optics coupling between the transmission phosphor layer and the intensifier, especially in HVEM operation, since lens coupling involves too large a light loss. Successful operation depends largely on optimization of the phosphor for the acceleration voltage used (mass thickness of phosphor between 1 mg/cm^2 and 5 mg/cm^2) but some images and diffraction patterns require higher resolution (1024 lines) and higher-priced television intensifier equipment.

Both static and dynamic (time-lapsed) images are most conveniently stored on a storage tube and then retrieved on video disc or printed by a hard copy device. This arrangement enables single frames (1/30 s) or integrated frames to be selected and stored with the desired time lapse between each image.

D. Electron Diffraction (ED)

The advantages of the electron diffraction technique over other diffraction methods have been discussed (HUI and PARSONS, 1974). They may be summarized as follows (1) Specimens of only one or several membrane thicknesses (25 to several hundred Å) can be examined because of the strong interaction between the specimen and the incident electron beam. (2) The data collection time is very short (of the order of seconds). (3) Very small areas can be observed by use of the selective area diffraction technique, and localized structure, such as domain structure, may be detected. The signal-to-noise ratio may also be improved by this method. (4) A very small amount of specimen is required and impurities can be screened. (5) The shape and morphology of the specimen may be observed by imaging. (6) The reflections of the diffraction pattern may be phased by the imaging method. The phases are derived either by direct comparison between the electron diffraction pattern and the Fourier transform of the image (UNWIN and HENDERSON, 1975), or by the GERCHBERG (1972) iteration method.

I. Selective Area Electron Diffraction

Selective area ED is normally achieved by introducing a selective area aperture in the image plane of the objective lens. The selective area may be restricted to some microns across, without interfering with the diffraction pattern, if the aperture is placed very close to the image plane. In this case, neighboring areas around the selected diffraction area are illuminated unnecessarily and some of the radiation-sensitive specimen is wasted. Alternatively, selective area ED may be done by restricting the illumination area to a few micrometers in diameter, using a pointed filament and a 10-μm second condenser aperture (HUI and PARSONS, 1975). Recently, TONOMURA (1974) used a field emission source to further limit the illumination area to the uncertainty principle limit of 400 Å in diameter. He is able to obtain an ED pattern from a small area (about 0.5 μm) of a fixed and stained section of tortoise retina membranes. With the development of lanthanum hexaboride filaments (BROERS, 1975) and field emission guns, small selective diffraction areas will be more readily available in commercial instruments.

By applying selective area ED, a domain structure of phospholipid bilayer was proposed (HUI et al., 1974). With an observation area (SAD) of 30 μm^2, the typical diffraction pattern is an arcked hexagonal pattern (Fig. 6 A), or several superimposed hexagonal patterns (Fig. 6 B). These patterns are superimposed on a sharp diffraction ring of the same spacing. On widening the SAD, more superimposed hexagonal patterns are recorded, and at SAD=800 μm^2 or greater, the pattern becomes an even ring powder pattern similar to those obtained by X-ray diffraction. Prolonged exposure or high

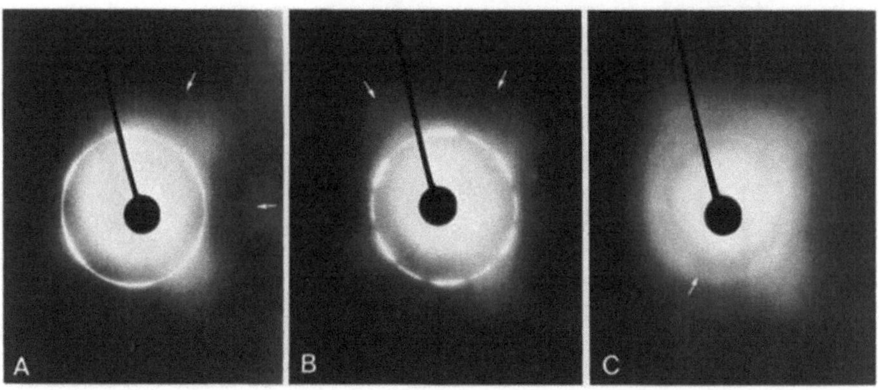

Fig. 6. A – C. ED patterns of dipalmitoyl lecithin bilayers taken below the transition temperature (*left* and *center*) with *arrows* indicating weak reflections, and above the transition temperature (*right*) (permission of American Association for the Advancement of Science)

beam intensity also reduces the hexagonal pattern to a uniform ring before the membrane is broken as a result of the thermal and radiation damage of the electron beam. Because X-ray diffraction shows only ring patterns due to its averaging nature, the existence of single crystal domains of some 5 μm diameter were revealed for the first time in this experiment. The spacings and intensities agree well with those calculated from the atomic scattering factors of hexagonally packed hydrocarbon chains. Based on this observation, a model of domains formed by a large number of small, but well-packed areas of crystallites, each slightly misoriented from the neighboring ones, was proposed (Fig. 7). The optical diffraction from this model is very similar to the ED pattern.

Below the transition temperature (T_c), the diffraction patterns of the domain appear to be static. This observation does not exclude the drifting (of one domain width) and the rotating (within the half arcking angle) of the domains during the photographic time of 5 s or so. The small angular oscillation of the crystallites within the half arcking angle is also possible. To measure this oscillation, a more sensitive electron detection system and faster re-

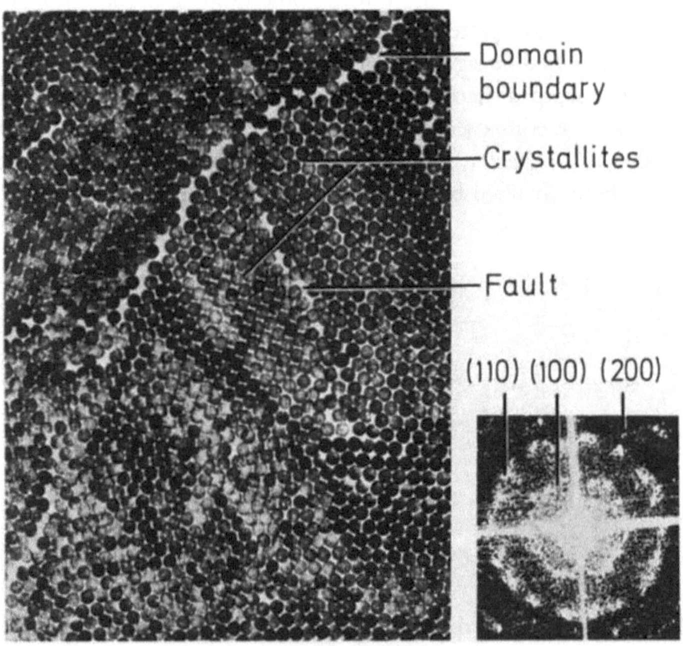

Fig. 7. A model for the packing of the hydrocarbon chains in phospholipid bilayers. A boundary separating two domains and faults separating slightly misoriented crystallites are shown. Pattern at the *right hand corner* is the optical transform of the same figure

cord method are needed. Recently a direct measurement of the membrane movement by electron diffraction and microscopy has been made, using an environmental chamber. Lateral diffusion coefficients have been derived from these measurements (HUI, 1976 b).

Most wet biological membranes show a diffuse diffraction band of spacing 4.7 Å. In certain cases, a weak ring at a spacing of 4.2 Å could be seen at low temperature. Electron diffraction from wet human erythrocyte ghost membrane has been reported previously (HUI and PARSONS, 1974). Most high angle reflections obtained from dehydrated membranes, such as that shown in Figure 1, do not appear when the specimen is wet.

II. Small-angle Electron Diffraction

High-resolution, small-angle ED may be achieved by one of several methods (FERRIER, 1969). In this respect ED has an advantage over XD because of the easily variable camera length. However, the unknown amount of small-angle inelastic electrons centered around the unscattered beam introduces many experimental and interpretational problems that are less prominent in X-ray diffraction. This inelastic background must be removed to recover weak signals from the scattering by irregular molecular arrangements.

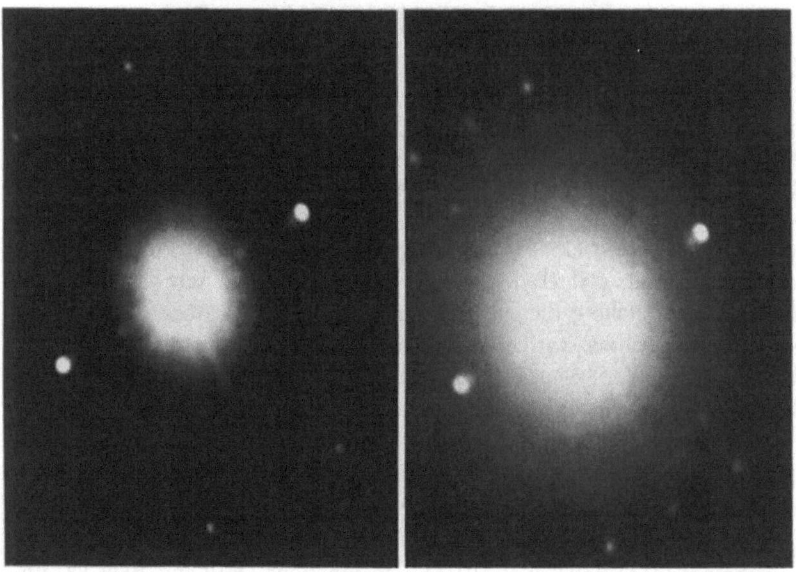

Fig. 8. Filtered (*left*) and unfiltered small angle electron diffraction pattern of a catalase crystal

Energy filtering helps to improve the situation. An example of energy filtering is shown in Figure 8 where a catalase microcrystal diffraction pattern is shown with and without filter (Hui and Ottensmeyer, unpublished).

Periodic structure of large spacing is not common in biomembranes, but there are several cases where these structures are known to exist. The purple membranes of halobacterium in water give the small-angle diffraction pattern shown in Figure 9. Up to 25 discrete rings were observed, the spacing rang-

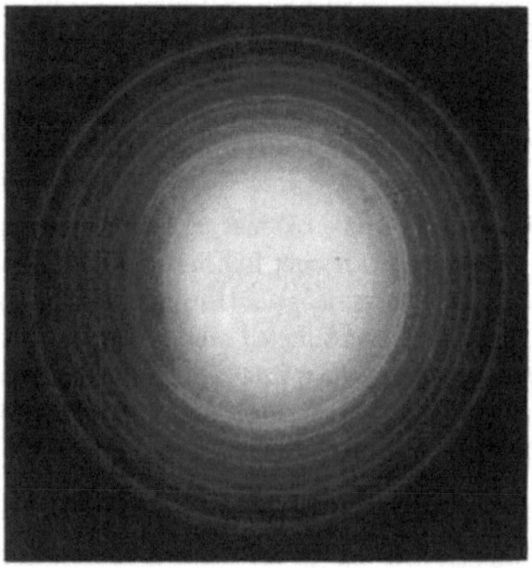

Fig. 9. ED pattern of hydrated purple membrane of *halobacterium halobium*

ing from 54 to 3.5 Å. All reflections may be accounted for if a hexagonal packing of rhodopsin molecules is assumed. A single-crystal pattern has been obtained from single discs of native and reconstituted purple membranes (Unwin and Henderson, 1975). The ED intensity data agree with X-ray diffraction results (Henderson, 1975). The reflections were phased by the image method, taking the lens transfer function into account. The unit cell structure was Fourier-synthesized from the electron diffraction data. This way Unwin and Henderson (1975) were able to solve the structure of a membrane-bound protein – bacteria rhodopsin – to a resolution of 7 Å.

Much work has been done on the optical diffraction of negative-stain or freeze-fracture micrographs of membranes in order to obtain information about the regular packing in a small area. Optical transform of micrographs of purple membranes (Blaurock and Stoeckenius, 1971), cytochrome oxidase membranes (Maniloff et al., 1973), gap junctions (Hui and Berger, unpublished), and retina rod disc (Blasie et al., 1969) have been reported. This information should be obtainable directly with selective area small-angle ED and with adequate electron energy filtering, thus avoiding the electron imaging process. However, there are certain technical problems, such as the radiation damage to the specimen and the lack of contrast in the specimens, that still need to be worked out. It is also interesting to note that the optical transform of freeze-fracture micrographs of human erythrocyte ghost membranes gives no detectable information on the distribution of the IMPs (HUI and BERGER, unpublished). It may be that the distribution of IMPs is more random than in cases in which nearest neighbor peaks are seen. This may explain the failure of XD and ED to verify the existence of IMPs in a natural state. Another possibility is that IMPs are in sufficient motion to give averaged scattering in XD and ED. If this is the case, one might expect prior aldehyde fixation to enable their presence to be revealed as diffraction rings.

Lamellar repeat of membranes also gives small-angle diffraction. Such cases are well known in X-ray diffraction for both natural membrane lamellae or artificially stacked ones. Most membrane lamellae are too thick for the 100 kV electron beam to penetrate along the lamellar face. HVEM opens up new possibilities of obtaining ED from membrane lamellae so as to enable structure profiles across the membrane to be deduced. Electron diffraction from the membrane lamellae of a single retina rod outer segment has been reported using an 800 kV beam (HUI et al., 1973). ED patterns may be obtained from a small fragment of lamella; therefore a large concentration of lamellar membranes, as required for XD is not necessary. It is also possible to select only a few repeat units from a membrane stack so that the quasi-continuous diffraction intensity may be analyzed by the Q-function method (HOSEMANN and BAGCHI, 1962).

III. Phase Transition and Phase Separation in Membranes

ED provides a unique way to study the phase characteristics of natural and reconstituted membranes in the form of a single membrane sheet as against artificially pelleted membrane stacks or concentrated suspensions. The structure of a single layer of membrane has been shown to be different from that in a multilayer state (HUI, 1976 a). In the latter state some lipid acyl chains may be tilted with respect to the bilayer surface, resulting in a different ED pattern. The tilt angle affects the calculation of area per molecule of the membrane.

With a temperature-controllable environmental chamber (HUI et al., 1976) one can study the phase diagram of a single bilayer membrane as functions of percentage humidity and temperature. It was found that the bilayer has a least-energy-packing spacing independent from the surface pressure of its 'precursor' monolayer (HUI et al., 1975). In DPPC bilayers, the spacings of the bilayer expand slightly with temperature until the sharp patterns disappear and a diffuse ring of spacing approximately 4.4 Å is obtained (Fig. 6 c). This indicates that the structure of the bilayer has passed through a phase transition from hexagonal packing to a liquid-crystalline structure. For all fully hydrated DPPC bilayers, this transition takes place abruptly at 41.5° C. For dried bilayers, the phase transition is still not seen at a temperature as high as 62° C. This result corresponds with the phase diagram obtained by X-ray diffraction of a mixed lipid/water system (CHAPMAN et al., 1967), showing a similarity, in this case, between the bulk system and a single bilayer (Fig. 10). Phase diagrams of multicomponent systems such as lecithin – cholesterol mixtures are more complicated. If thermally induced phase

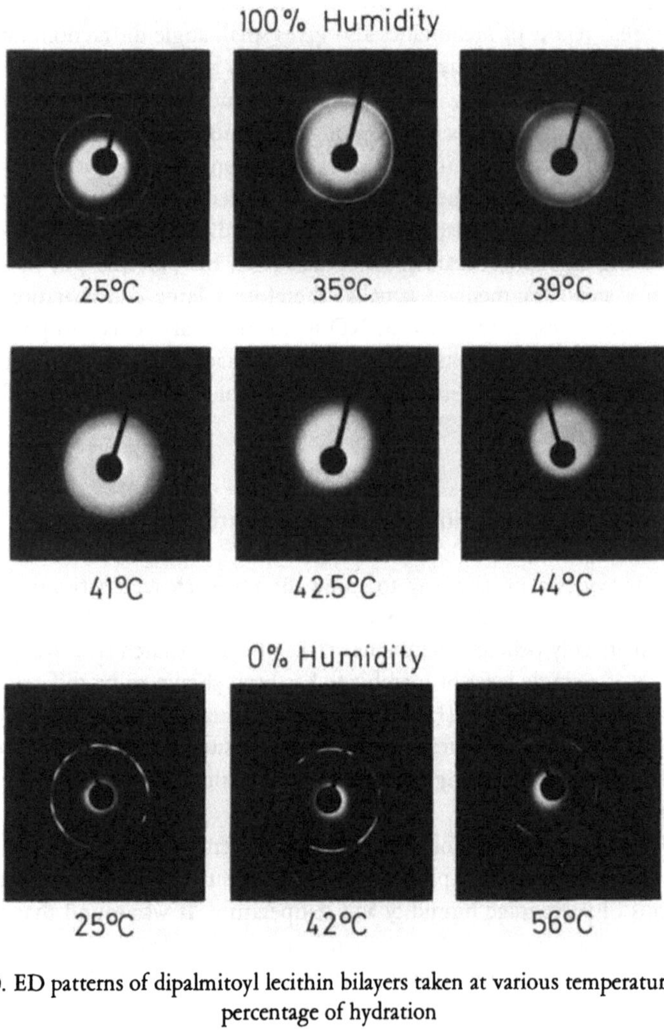

Fig. 10. ED patterns of dipalmitoyl lecithin bilayers taken at various temperatures and percentage of hydration

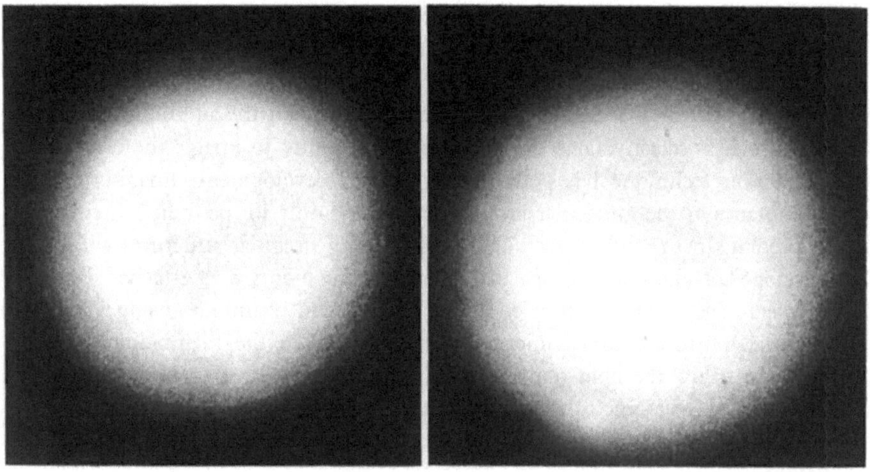

Fig. 11. ED patterns of the plasma membranes of rat hepatoma cells at 15° C (*left*) and 18° C. The sharp ring at a spacing of 4.1 Å seen at 15° is no longer observable at 18° C (permission of American Association for Cancer Research)

separation occurs, there will be superposition of several types of diffraction patterns. Selective area ED may give the size of each phase-separated domain.

This kind of mixed pattern also appears in the ED of natural membranes. The transition temperature of any of the phases, if measurable, may give some information about the constituents and structure of the membrane. This measurement has been used to characterize cancer cell membranes in relation to their normal counterparts. Figure 11 shows ED patterns of the plasma membrane from Reuber H-35 hepatoma at two temperatures. The change of transition temperature with storage time at 4° C under various conditions for rat hepatoma and hepatocyte membranes differs. Oxidation was found to be a cause of the change (HUI and PARSONS, 1976).

E. Conclusions and Future Development

Given sufficient contrast and radiation tolerance, the hydrated, unfixed, and unstained membrane specimen should give more detailed and less artifactual information than those prepared by conventional methods (HUI and PARSONS, in press; UNWIN and HENDERSON, 1975). Regardless of the freezing, substitution, or hydration chamber method used, the same problems of radiation damage and lack of contrast are encountered. These difficulties have been somewhat overcome by various techniques discussed in this Chapter. In the hydration chamber method, the specimens are least affected by artifacts.

Since they are under physiologic environment, *in situ* experiments may be performed. At the present stage of development, the resolution of ED data on hydrated membranes is equal to those of XD data and has not been exceeded by any freezing or embedding EM technique. With the additional advantages such as selective area diffraction and the ability to image, the electron diffraction technique has potential for further development. Imaging wet membranes at the molecular level has been done only for periodic structures at present. In the future, techniques for imaging irregular structures will be developed. The mobility of isolated membrane fragments with attached dense markers (e.g., ferritin) can be studied by our present techniques using a temperature-controlled environmental chamber to place the membrane either above or below the lipid transition temperature. Time-lapse images obtained on a television storage tube system should give quantitative data on the lateral diffusion constants of individual domains of natural and synthetic membranes.

The ideal is, of course, to study the mobility, chemical activity, and structure of the membranes of cells in the living state. While many 'probe' experiments, e.g., with fluorescent-coupled surface markers, are possible, it is altogether a much larger task to image living cell membranes. Perhaps the most attractive new development is that of ROBINSON (1975) using backscatter images of wet biological objects. We are pursuing this possibility using a differentially pumped chamber fitted to a standard scanning microscope. Fixed-beam transmission experiments with living cells may eventually prove possible in the 1 MeV microscope with a hydration chamber. So far, no definite evidence of cell movement has been obtained. The reasons are complex even when radiation damage is reduced to a reasonable level. It appears that cell cytoplasm motion (and even Brownian motion of inorganic sols) can become frozen under certain environmental, substrate, and electron beam conditions. The situation awaits further clarification.

Acknowledgements. The work by the authors was supported by USPHS grants GM16454 and CA15330 and in part by USPHS grants CA13845, RR00754, and American Cancer Society Grant BC-248, NIH-HVEM facilities, contract DRR-70-4136 to U. S. Steel and NSF grant GB38346. Collaboration with Drs. F. P. OTTENSMEYER, D. PAPAHADJOPOULOS, and J. BERGER is appreciated. S. W. HUI is a recipient of NIH Career Development Award CA00084.

References

Allinson, D. F., Gosnold, A. W. O., Loveday, M. S.: Modified environmental cell for use in a high voltage electron microscope. In: Proc. 5th Europ. Cong. Elect. Microscopy, Oxford Univ. Press, 1972, pp. 336 – 337

Basu, S., Parsons, D. F.: Replication of wet objects. In: Proc. 8th. Intern. Cong. EM, II. Australian Acad. Sci., Canberra, Australia. Sanders, J. V., Goodchild, D. J. (eds.). 1974, pp. 168 – 169

Bernhard, W.: Progress in cyro-ultramicrotomy. In: Proc. 8th Intern. Cong. EM, II. Australian Acad. Sci., Canberra, Australia. 1974, pp. 706 – 707

Blasie, J. K., Worthington, D. R., Dewey, M. M.: Molecular localization of frog retina receptor photopigment by electron microscopy and low angle X-ray diffraction. J. Mol. Biol. **39**, 79 – 94 (1969)

Blaurock, A. E., Stoeckenius, W.: Structure of the purple membrane. Nature (New Biol.) **253**, 152 – 155 (1971)

Broers, A. N.: Electron sources for scanning electron microscopy. In: Scanning Electron Microscopy/1975. O. Johari (ed.). Chicago: IIT Res. Inst., 1975, pp. 661 – 670

Carlemalm, E., Weislander, A.: E. D. studies of biological membranes and lipids. Nature (London) **254**, 537 – 538 (1975)

Chapman, D., Williams, R. M., Ladbrooke, B. P.: Physical studies of phospholipids. Chem. Phys. Lipids **1**, 445 – 475 (1967)

Christensen, A. K.: Frozen thin sections of fresh tissue for electron microsc. with a description of pancreas and liver. J. Cell Biol. **51**, 772 – 804 (1971)

El Hili, A.: Analyse quantitative a haute resolution par images electroniques filtrees. J. Microscopie **5**, 669 – 690 (1966)

Emmelot, P., Bos, C. J., Benedetti, E. L., Rumke, P. H.: Studies on plasma membranes, Biochim. Biophys. Acta **90**, 126 – 145 (1964)

Fernandez-Moran, H.: Diffraction of electrons by structure resembling myelin lamellae. Exptl. Cell Res. **2**, 673 – 679 (1951)

Ferrier, R. P.: Small angle electron diffraction in the electron microscope. Adv. Opt. Electron Micros. **3**, 155 – 218 (1969)

Fromberg, P.: Electron microscope studies of lipid protein film. Nature (London) **231**, 267 – 268 (1971)

Gerchberg, R. W.: Holography without fringes in electron microscopy. Nature (London) **240**, 404 – 406 (1972)

Heide, G. G., Grund, S.: Wet biological specimens in the electron microscope. Transfer and observation at low temperature. In: Proc. 8th Intern. Cong. EM, II. Australian Acad. Sci., Canberra, Australia. Sanders, J. V., Goodchild, D. J. (eds). 1974, pp. 46 – 47

Henderson, R.: The structure of the purple membrane from *Halobacterium halobium*. J. Mol. Biol. **93**, 123 – 138 (1975)

Henkelman, R. M., Ottensmeyer, F. P.: Macromolecular fine structures by dark field electron microscopy. Biophys. J. **9**, 11 – 44 (1969)

Henkelman, R. M., Ottensmeyer, F. P.: An energy filter for biological electron microscopy. J. Microscopy **102**, 79 – 94 (1974)

Hui, S. W.: The tilting of the hydrocarbon chains in a single bilayer of phospholipid. Chem. Phys. Lipids, **16**, 9 – 18 (1976 a)

Hui, S. W.: Direct measurement of membrane motion and fluidity by electron microscopy. Nature (London) **262**, 303 – 305 (1976 b)

Hui, S. W., Cowden, M., Papahadjopoulos, D., Parsons, D. F.: Electron diffraction study of hydrated phospholipid single bilayers: Effect of temperature, hydration &

surface pressure of the precursor monolayer. Biochim. Biophys. Acta **382,** 265 – 275 (1975)

Hui, S. W., Hausner, G. G., Parsons, D. F.: A temperature-controllable hydration or environmental stage for Siemens Elmiskop Ia. J. Phys. E.: Sci. Instrum. **9,** 69 – 72 (1976)

Hui, S. W., Parsons, D. F.: Electron diffraction of wet biological membranes. Science **184,** 77 – 78 (1974)

Hui, S. W., Parsons, D. F.: Direct observation of domains in wet lipid bilayers. Science **190,** 383 – 384 (1975)

Hui, S. W., Parsons, D. F.: Electron microscopy and electron diffraction of biological materials. In: Electron Microscopy. Barnett, R. J., Higgins, J. A. (eds.). New York: Wiley Interscience, in press.

Hui, S. W., Parsons, D. F.: Phase Transition in the Plasma Membranes of Rat Hepatocyte and Hepatoma Cells by Electron Diffraction. Cancer Res. **36,** 1918 – 1922 (1976)

Hui, S. W., Parsons, D. F., Cowden, M.: ED of wet phospholipid bilayers. Proc. Natl. Acad. Sci. U. S. **71,** 5968 – 5972 (1974)

Hui, S. W., Parsons, D. F., Matricardi, V. R.: Electron diffraction of wet retina rod disc membranes. In: Proc. 31st EMSA. Baton Rouge, LA: Claitors Publishing Division. C. J. Arceneaux (ed.) 1973, pp. 602 – 603

Hosemann, R. O., Bagchi, R. O.: Direct Analysis of Diffraction by Matter. Amsterdam: North Holland, 1962

Hurst, H.: An electron diffraction study and orientation of the lipids in yeast and bacterial cell wall. J. Exp. Biol. **29,** 30 – 53 (1952)

Joy, R. T.: Electron microscopical observation of aqueous biological specimens. Adv. Opt. Electron Micros. **5,** 297 – 352 (1973)

Khase, R. S., Misra, K. K.: Electron diffraction in the effect of certain drugs on cell membranes. Studia Biophysica **38,** 205 – 209 (1973)

Maniloff, J., Venderkooi, G., Hayashi, H., Capaldi, R. A.: Optical analysis of electron micrographs of cytochrome oxidase membranes. Biochim. Biophys. Acta **298,** 180 – 183 (1973)

Matheja, J. H.: ED of Membranes. Biophysik **7,** 163 – 168 (1971)

Matricardi, B. R., Moretz, R. C., Parsons, D. F.: Electron diffraction of wet proteins-catalase. Science **177,** 268 – 270 (1972 a)

Matricardi, V. R., Wray, G. P., Parsons, D. F.: Evaluation of emulsions and other recording media for 100 kV and 1000 kV electron microscopes. Micron **3,** 526 – 530 (1972 b)

Moretz, R. C., Akers, C. K., Parsons, D. F.: Use of small angle X-ray diffraction to investigate disordering of membranes during preparation for electron microscopy. I. Osmium tetroxide and potassium permanganate. Biochim. Biophys. Acta **193,** 1 – 11 (1969). II. Aldehydes, Biochim. Biophys. Acta **193,** 12 – 21 (1969)

Parsons, D. F.: Structure of wet specimens in electron microscopy. Science **186,** 407 – 414 (1974)

Parsons, D. F., Lyon, N. C.: Design of SEM hydration chambers and possibilities for viewing live biological material. In: Scanning Electron Microscopy/1975. Chicago: IIT Research Institute. O. Johari (ed.). 1975, pp. 28 – 34

Parsons, D. F., Matricardi, V. R., Moretz, R. C., Turner, J. N.: Electron microscopy and diffraction of wet, unstained and unfixed biological objects. Adv. Biol. Med. Phys. **15**, 162 – 270 (1974)

Parsons, D. F., Matricardi, V. R., Subjeck, J., Uydess, I., Wray, G.: High voltage electron microscopy of wet whole cancer and normal cells. Biochim. Biophys. Acta **290**, 110 – 124 (1972)

Parsons, D. F., Subjeck, J. R.: The morphology of the polysaccharide coat of mammalian cells. Biochim. Biophys. Acta, Reviews on Membranes **265**, 85 – 113 (1972)

Pease, D. C.: Substitution techniques. In: Advanced Techniques in Biological Electron Microscopy. Koehler, J. K. (ed.). Berlin-Heidelberg-New York: Springer, 1973, pp. 35 – 63

Pease, D. C.: Polar embedments which retain lipids in sections. In: Proc. 8th Intern. Cong. EM, II. Sanders, J. V., Goodchild, D. J. (eds.) 1974, pp. 126 – 127

Robertson, J. G., Parsons, D. F.: Myelin structure and retention of cholesterol in frog sciatic nerve embedded in a resorcinol formaldehyde resin. Biochim. Biophys. Acta **219**, 379 – 387 (1970)

Robinson, V. N. E.: Backscattered electron imaging. IN: Scanning Electron Microscopy/1975. O. Johari (ed.). Chicago: ITT Research Institute, 1975, pp. 51 – 60

Robinson, V. N. E.: A wet stage modification to a scanning electron microscope. J. Microscopy **103**, 71 – 77 (1976)

Rose, H., Plies, E.: Design of a magnetic energy analyzer with small aberrations. Optik **40**, 336 – 341 (1974)

Stoeckenius, W., Rowen, R.: A morphological study of *Halobacterium halobium* and its lysis in media of low salt concentration. J. Cell Biol. **34**, 365 – 393 (1967)

Swann, P. R.: High voltage studies of environmental reactions. In: Electron Microscopy and Structure of Materials. Thomas, G., Fulrath, R. M., Fisher, R. M. (eds.). Berkeley-London: Univ. Calif. Press, 1972, pp. 878 – 904

Taylor, K. A., Glaeser, R. M.: Electron diffraction of frozen, hydrated protein crystals. Science **186**, 1036 – 1037 (1974)

Tonomura, A.: High resolution electron diffraction from a small object in a field emission electron microscope. Optik **39**, 386 – 395 (1974)

Turner, J. N., Hausner, G. G., Parsons, D. F.: An optimized Faraday cage design for electron beam current measurement. J. Phys. E.: Sci. Instrum. **8**, 954 – 957 (1975)

Unwin, P. N. T., Henderson, R.: Molecular structure determination by electron microscopy of unstained crystalline specimens. J. Mol. Biol. **94**, 425 – 440 (1975)

Vail, W. J., Riley, R. K.: The structure of cytochrome oxidase membranes. FEBS. Lett. **40**, 269 – 273 (1974)

Ververgaert, P. H. J., Verkleij, A. J., Elbers, P. F., van Deenan, L. L. M.: Analysis of the crystallization process in lecithin liposomes: Freeze etch study. Biochim. Biophys. Acta **311**, 320 – 329 (1973)

Waldbillig, R. C., Robertson, J. D., McIntosh, T. J.: Images of divalent cations in unstained symmetric and asymmetric lipid bilayers. Biochim. Biophys. Acta **448**, 1 – 14 (1976)

Warley, A., Cook, G. M. W.: Plasma membranes of lymphoid Cells. Biochim. Biophys. Acta **323**, 55 – 68 (1973)

Subject Index

Advanced Techniques in Biological Electron Microscopy

Editor: J. K. Koehler

With contributions by S. Bullivant, J. Frank, K. Hama, T. L. Hayes, J. H. Luft, F. A. McHenry, D. C. Pease, M. M. Salpeter

108 figures. XII, 304 pages. 1973
ISBN 3-540-06049-9

This book will serve not only as a reference work for scientists already expert in biological electron microscopy, but more importantly, will stimulate biologists to investigate and employ techniques which they may have considered too exotic or complex.

Advanced electron optical and preparatory techniques are treated thoroughly. Discussions of new embedding techniques such as inert dehydration, freeze-fracturing, and autoradiography are included. The emphasis is on analytical methodology and interpretation. Other chapters deal with image processing using computers, scanning electron microscopy, and high voltage electron microscopy. In a number of instances, including the chapters on embedding media and computer processing, material not previously available to a general audience is presented in detail.

Contents

Embedding Media–Old and New. – Substitution Techniques. – Freeze-Etching and Freeze-Fracturing. – Electron Microscope Autoradiography. – Scanning Electron Microscope Techniques in Biology. – Computer Processing of Electron Micrographs. – High Voltage Electron Microscopy. – References. – Subject Index.

Springer-Verlag
Berlin
Heidelberg
New York

MIX
Papier aus verantwortungsvollen Quellen
Paper from responsible sources
FSC® C105338

If you have any concerns about our products,
you can contact us on
ProductSafety@springernature.com

In case Publisher is established outside the EU,
the EU authorized representative is:
Springer Nature Customer Service Center GmbH
Europaplatz 3, 69115 Heidelberg, Germany

Printed by Libri Plureos GmbH
in Hamburg, Germany